RADIOACTIVE WASTE MANAGEMENT IN THE 21ST CENTURY

RADIOACTIVE WASTE MANAGEMENT IN THE 21ST CENTURY

William R Roy
University of Illinois at Urbana-Champaign, USA

World Scientific

EW JERSEY • LONDON • SINGAPORE • BEIJING • SHANGHAI • HONG KONG • TAIPEI • CHENNAI • TOKYO

Published by

World Scientific Publishing Co. Pte. Ltd.
5 Toh Tuck Link, Singapore 596224
USA office: 27 Warren Street, Suite 401-402, Hackensack, NJ 07601
UK office: 57 Shelton Street, Covent Garden, London WC2H 9HE

Library of Congress Cataloging-in-Publication Data
Names: Roy, William R., 1954– author.
Title: Radioactive waste management in the 21st century / William R. Roy, University of Illinois at Urbana-Champaign, USA.
Description: New Jersey : World Scientific, [2021] | Includes bibliographical references and index.
Identifiers: LCCN 2021026277 | ISBN 9789811228292 (hardcover) | ISBN 9789811230080 (paperback) | ISBN 9789811228308 (ebook for institutions) | ISBN 9789811228315 (ebook for individuals)
Subjects: LCSH: Radioactive waste disposal.
Classification: LCC TD898.17 .R69 2021 | DDC 621.48/38--dc23
LC record available at https://lccn.loc.gov/2021026277

British Library Cataloguing-in-Publication Data
A catalogue record for this book is available from the British Library.

For any available supplementary material, please visit
https://www.worldscientific.com/worldscibooks/10.1142/12042#t=suppl

Desk Editor: Amanda Yun

Typeset by Stallion Press
Email: enquiries@stallionpress.com

This textbook is dedicated to my father
William Windsor Roy, Industrial Engineer

Preface

"He had no idea of the journey we were about to take."

— Henry Lawson in *Journey to the Center of the Earth* by Jules Verne

Anti-nuclear activists claim that the world should abandon nuclear energy because we do not know how to safely manage radioactive wastes, and that previous attempts to manage *even* low-level wastes have been environmental disasters. Unfortunately, this biased and distorted depiction is believed by many people and has spread paranoia and an irrational fear and public distrust about *anything* radioactive. This fear has done little to persuade governments in some countries to make policy decisions that are based on the recommendations of scientists and engineers about managing radioactive wastes. Instead, the decisions in some countries are based on emotions and politics, or simply left unresolved for years (Roy, 2013).

As the title implies, this textbook is about the management of radioactive wastes generated by nuclear energy production, nuclear weapons production, military applications, industrial testing and fabrication, and medical applications of radiation sources. In terms of energy production, wastes result from the routine operation of nuclear power plants, the decommissioning and demolition of former power plants and research reactors, and nuclear fuel fabrication and reprocessing facilities. The future of nuclear energy varies between countries as discussed in Chapter 10. The status of nuclear energy in the United States (US), Canada and the Republic of Finland, for example, is stable. The Federal Republic of Germany and the Republic of China are phasing nuclear energy out. The German public opposes nuclear energy in part because of the government's dismal attempt to manage radioactive wastes in the past. The Russian Federation and the People's Republic of China seek to expand the role of nuclear energy. The prevalent method for disposing of most radioactive wastes is by shallow-land burial. Wastes such as used nuclear fuel from power plants, however, need to be placed in long-term storage in deep-geological repositories (the topic of Chapter 6). Used nuclear fuel will remain radioactive for thousands of years. Regardless of the future of nuclear energy in the world, these wastes *already* exist world-wide, and require careful management *now*.

This goal of this textbook is to present concise, practical and updated material about the management of radioactive wastes. No single textbook on this topic can possibly cover all the knowledge available to the reader. The author has read several previous textbooks for content, style and pleasure. Some of these textbooks are summarized at the end of this Preface for the benefit of the reader. This textbook was written with an emphasis on geology and chemistry. An emphasis on international waste management was also made because it seems to be absent in other textbooks. The intended audiences for this textbook are undergraduate and graduate students in science and engineering, college-level faculty, attendees at waste-management training courses, Federal and commercial waste management/site remediation staff, environmental consultants, and waste management personnel in the military.

This was not an easy textbook to write because many aspects about the management of radioactive waste are in a state of flux. The current status of the proposed Yucca Mountain Nuclear Waste Repository in the US is unclear because of never-ending litigation. Similarly, the Federal Republic of Germany currently has no definite plans for the disposal of Heat-Generating Wastes (defined in Chapter 10). Radioactive waste management in Germany is currently plagued by bureaucratic governmental agencies that change with each election. Meanwhile, the Kingdom of Sweden and the Republic of Norway are leading the way to completion of a geological repository.

Before discussing waste management in the 21st century, it is instructive to briefly consider how far we have come in managing radioactive wastes. Starting in the mid-20th century, early attempts to manage low-level radioactive wastes (as discussed in Chapter 4) were tainted by a lack of experience and a regulatory framework. In the US, incomplete site investigations, poor public relations, a lack of waste compaction and sorting, the use of easily degradable cardboard and fiberboard boxes, waste transportation blunders, and trench-cover failures among others led to the closure of most early near-surface disposal sites. What have we learned since then?

Early waste management at the Hanford Nuclear Reservation in the State of Washington (US) is a prime example of the mismanagement of radioactive wastes. Hanford is part of the former Nuclear Weapons Complex. During World War II and the Cold War, Hanford produced plutonium for nuclear weapons. The operations and waste-management practices at Hanford resulted in soil and groundwater contamination on an unprecedented scale. Hanford relied on leaking storage tanks, and intentionally released radioactive and chemical wastes into ponds, trenches, and injection wells. For example, one waste management "technology" was called a crib. A crib was simply a ditch. It was described by Saddington and Templeton (1958) which may be the oldest known book published on radioactive waste management. A crib was typically 6.1 m deep and 427 m long and backfilled with gravel (Fig. P.1). Untreated liquid wastes generated by the production of plutonium were discharged into the crib to migrate downward (Fig. P.2). It was assumed that the radionuclides dissolved in the waste liquids would be immobilized by ion exchange associated with the clay minerals. It was acknowledged that tritium, ruthenium, and nitrate would not be immobilized, and could "flow to groundwater" but "be rapidly diluted." Both ion exchange and groundwater dilution have limits in their ability to reduce the impacts of groundwater contamination. The volumes of wastes added to the cribs *far* exceeded those limits. Hanford is now the largest and most expensive environmental cleanup project in the US.

Bradley (2017) provided a detailed account about how radioactive wastes were mismanaged in the former USSR. In some cases,

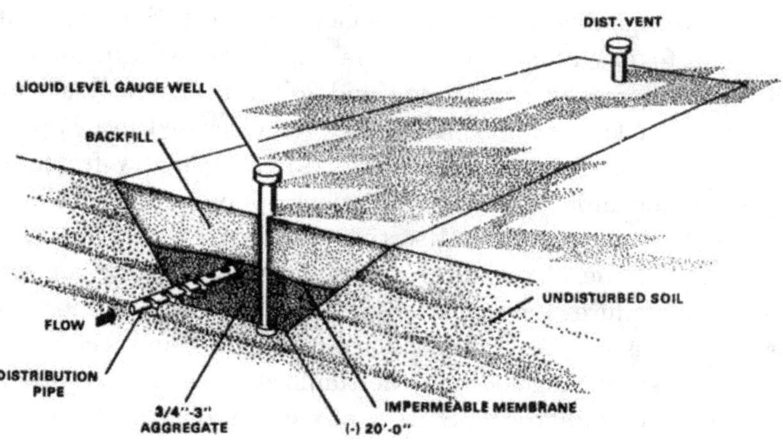

Figure P.1. A crib for liquid radioactive waste disposal at the Hanford Nuclear Reservation in the US (Saddington and Templeton, 1958).

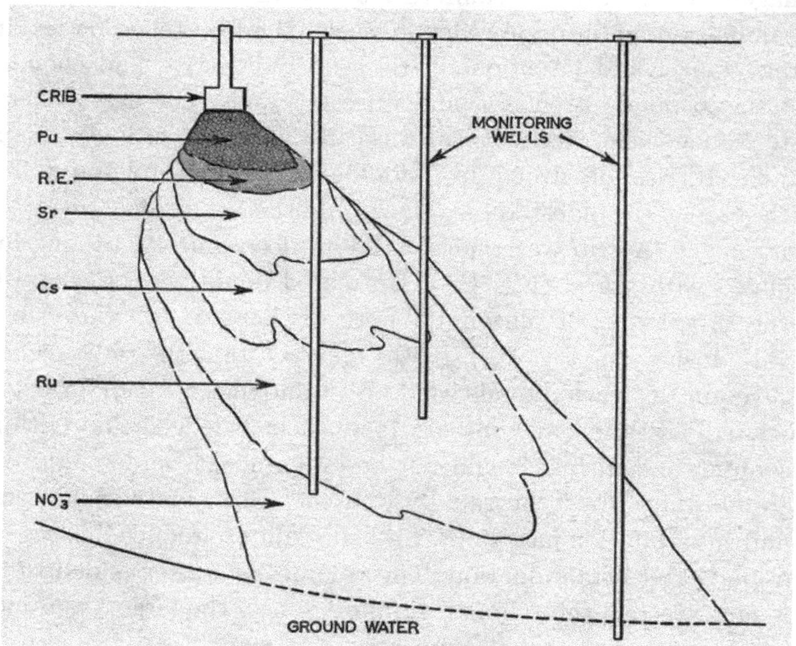

Figure P.2. The downward migration of radionuclides and nitrate from a crib at the Hanford Nuclear Reservation as envisioned by Saddington and Templeton (1958). R.E. refers to rare earths.

there was little distinction between radioactive wastes generated by the Russian Weapons Complex from the Cold War and electric utilities, and they were co-disposed in rivers and ponds resulting a widespread groundwater contamination. Trucks transporting radioactive wastes along the roads of Moscow often spilled wastes prior to about 1990. A rod of cobalt-60 was found in a children's play area with an activity of 40 mSv/h (4 rem/h). Bradley also recorded that a 6.12 mSv/h (612 mrem/h) source was found in a sandbox at a daycare center in Moscow. Are these activities a concern? That is the subject of Chapter 1.

This textbook started as simple lecture notes and class handouts for a course at the University of Illinois at Urbana-Champaign. This course is NPRE 442, *Radioactive Waste Management* in the Department of Nuclear, Plasma and Radiological Engineering (NPRE). The author has taught this course since 2008. This textbook benefited from the author's long career of conducting research in the laboratory and in the field at the Illinois State Geological Survey in Urbana-Champaign. Waste management was a common theme in many of the research projects, and included municipal wastes, coal-related wastes, deep-well injection of liquid hazardous wastes, municipal incinerator wastes, agricultural and urban wastes, geological sequestration of waste gases, and lastly of course radioactive wastes. An integral part of many of the projects was the geochemical fate and transport of inorganic and organic contaminants in air, geomedia, and groundwater. The author greatly benefited from teaching for the Royal Institute of Technology in Sweden in a course called SH262V, *Geological Storage of Nuclear Spent Fuel* since 2015. Draft versions of some of the chapters in this textbook were critically reviewed by former students in a course taught by the author since 2016 (NPRE/GLBL (Global Studies) 481, *Writing on Technology and Security*): Yaw Akpaloo, Shashikiran Duraisamy, Iti Mehta, Chinedu Oputa, and Aric Tate. Prof. Cliff Singer of NPRE also reviewed a draft chapter. My patient wife, Bobbi, also reviewed every draft chapter. With all that said, I hope that this textbook is valuable to all future students, faculty, and professionals who need to understand radioactive waste management.

William R. Roy
October 2020
Urbana, Illinois, USA

Bibliography

Bradley, D. J. (2017). *Behind the Nuclear Curtain: Radioactive Waste Management in the Former Soviet Union.* Battelle Press, Columbus, Ohio.

Roy, W. R. (2013). Radioactive Waste Management: What Does the Public Need to Know? *Journal of Nuclear Energy Science and Power Generation Technology.* Available t: http://dx.doi.org/10.4172/2325-9809.S1-005.

Saddington, K. and Templeton, W. L. (1958). *Disposal of Radioactive Waste.* George Newnes Limited, London.

Bibliography of Books about Radioactive Wastes Published After 1980

All waste types

Berlin, R. E. and Stanton, C. C. (1989). *Radioactive Waste Management.* John Wiley and Sons, Inc., New York, New York.

Brookins, D. G. (1984). *Geochemical Aspects of Radioactive Waste Disposal.* Springer-Verlag, Inc., New York, New York.

Chapman, N. A. and McKinley, I. G. (1987). *The Geology of Disposal of Nuclear Waste.* John Wiley and Sons, U.K.

Murray, R. L. (2003). *Understanding Radioactive Waste*, 5th Ed., Battelle Press, Columbus, Ohio.

Saling, J. H. and Fentiman, A. W. (2001). *Radioactive Waste Management*, 2nd Ed., Taylor and Francis, New York, New York.

Wiles, D. R. (2002). *The Chemistry of Nuclear Fuel Waste Disposal.* Polytechnic International Press, Montreal, Canada.

Spent nuclear fuel and high-level waste

National Research Council. (1983). *A Study of the Isolation System for Geological Disposal of Radioactive Wastes by the Waste Isolation Systems Panel.* National Academy Press, Washington, D.C.

Savage, D. (ed). (1995). *The Scientific and Regulatory Basis for the Geological Disposal of Radioactive Waste.* John Wiley & Sons, West Sussex, England.

Roxburgh, I. S. (1987). *Geology of High-Level Nuclear Waste Disposal: An Introduction.* Chapman and Hall, London, UK.

Reprocessing used nuclear fuel

Choppin, G. R., Khankhasayev, M. K., and Plendl, H. S. (2002). *Chemical Separations in Nuclear Waste Management.* Battelle Press, Columbus, Ohio.

Low-level radioactive waste

Weigart, J. (2007). *Waste is a Terrible Thing to Mind.* Rivergate Books, New Brunswick New Jersey.

Geochemistry of radioactive contamination

Siegel, M. D. and Bryan, C. R. (2003). Environmental Geochemistry of Radioactive Contamination. Sandia National Laboratories, Albuquerque, New Mexico (Report number SAND2003-2063).

Decommissioning and wastes

Rhaman, A. (2008). *Decommissioning and Radioactive Waste Management.* Whittles Publishing, Catherines, Scotland.

Transporting radioactive wastes

National Research Council. (2006). *Going the Distance?* The National Academies Press, Washington D.C.

Nuclear site restoration

National Research Council. (2001). *Science and Technology for Environmental Cleanup at Hanford.* The National Academies Press, Washington D.C.

Contents

Contents

Chapter 1

Radiation and Exposure

"During the whole of the next day we proceeded on our journey through this interminable gallery, arch after arch, tunnel after tunnel."

— Henry Lawson in *Journey to the Center of the Earth* by Jules Verne

1.1 Introduction

Knowledge about the sources of radioactivity, radiation dosimetry and health physics are essential for the management of radioactive wastes. Radioactivity is a natural physicochemical process in which unstable elements spontaneously decay in order to attain a more stable atomic arrangement. Radiation is the emission of electromagnetic waves of energy and subatomic particles during radioactive decay. In managing radioactive wastes, the major concern is ionizing radiation which has enough kinetic energy to remove electrons from atoms or molecules, thus ionizing them (ATSDR, 1999). The potential adverse impact on health from this phenomenon is that ionizing radiation damages genetic material such as DNA. The impacted cells may not repair themselves correctly and lead to cancers. Exposure refers to an event when any material is subjected to any type of radiation. The subject of this chapter, however, is the measurement of exposure to ionizing radiation and minimizing occupational exposures from radioactive wastes.

1.2 The Half-Life

The rate by which unstable elements decay can be described as a first-order reaction: the rate depends on the activity of only one reactant. The half-life of a radionuclide is the time for one-half of the initial activity to decay into a daughter product. The daughter product may be stable, or radioactive and will decay into another daughter product (Saling and Fentiman, 2002). Uranium, for example, decays into a long chain of daughter products (see EPRI, 2014). As time progresses, the initial activity is reduced by one-half, then one-fourth, one-eighth and so forth, as a sequence of $(1/2)^n$ where n is the number of half-lives. For practical purposes, a radionuclide is often considered insignificant after 10 half-lives in which 99.9% of the initial activity has been depleted.

The biological half-life is the time required for the summation of available biological processes to eliminate one-half of a radionuclide that has been retained by the human body. For example, after cesium-137 enters the human body, the kidneys begin to remove it from the blood, and some cesium is released from the body in urine. A small portion is also released in the feces. Some of the cesium-137 that the body absorbs can remain in the body for weeks or months, but it is slowly eliminated through the urine and feces (ATSDR, 2004). An estimated biological half-life of cesium-137 in adult males is 105 days (Ashraf *et al.*, 2014).

The effective half-life is the time required for the activity of a radionuclide to be halved as a result of both radioactive decay and biological elimination. Effective half-life is calculated as

$$T_{\text{eff}} = (t_{\text{bio}} \times t_{1/2})/(t_{\text{bio}} + t_{1/2}) \tag{1.1}$$

where T_{eff} is the effective half-life, t_{bio} is the biological half-life and $t_{1/2}$ is the radioactive half-life

In the example with cesium-137, the radioactive half-life is 30.2 years (11,023 days) (Audi *et al.*, 2003). The effective half-life in an adult male becomes

$$(105 \text{ days} \times 11,023 \text{ days})/105 \text{ days} + 11,023 \text{ days}) = 104 \text{ days}.$$

In this case, the relatively rapid biological elimination dominates the process.

1.3 Modes of Radioactive Decay

Radioactive decay produces alpha, beta, and gamma radiation. Alpha radiation is composed of two protons and two neutrons. An alpha particle (α) resembles a helium ion with both electrons removed (He^{2+}). An alpha particle has 4 atomic-mass units, and its emission will yield a new radionuclide. For example, polonuium-210 decays to lead-206 by alpha decay with a half-life of 138.4 days (Audi *et al.*, 2003).

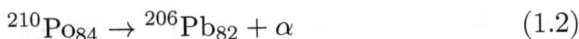

$$^{210}\mathrm{Po}_{84} \rightarrow {}^{206}\mathrm{Pb}_{82} + \alpha \tag{1.2}$$

Beta radiation (β) consists of the emission of electrons (e^-) or positrons (e^+). An unstable atomic nucleus with an excess number of neutrons may undergo β^- decay in which a neutron (η) is converted into a proton (ρ), an electron, and an antineutrino ($\bar{\upsilon}$):

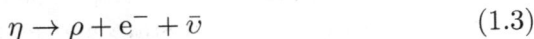

$$\eta \rightarrow \rho + \mathrm{e}^- + \bar{\upsilon} \tag{1.3}$$

An example of β^- decay is carbon-14 decay to nitrogen-14 with a half-life of 5,730 years (Audi *et al.*, 2003).

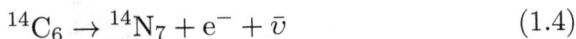

$$^{14}\mathrm{C}_6 \rightarrow {}^{14}\mathrm{N}_7 + \mathrm{e}^- + \bar{\upsilon} \tag{1.4}$$

During positron emission (β^+ decay), a proton is converted into a neutron, a positron and a neutrino (υ):

$$\rho \rightarrow \eta + \mathrm{e}^+ + \upsilon \tag{1.5}$$

An example of β^+ decay is the rubidium-82 decay to krypton-82 with a half-life of 76 s (Audi *et al.*, 2003)

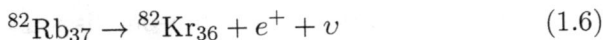

$$^{82}\mathrm{Rb}_{37} \rightarrow {}^{82}\mathrm{Kr}_{36} + e^+ + \upsilon \tag{1.6}$$

Gamma radiation (γ) consists of photons that are emitted from an unstable nucleus, typically after alpha or beta decay. X-rays are a form of electromagnetic radiation but emanate from electron transitions between atomic orbits (ATSDR, 1999). An example of gamma emission is the β^- decay of the neutron-activated product cobalt-60 which decays to nickel-60 with a half-life of 5.27 years (Unterweger *et al.*, 2016)

$$^{60}\mathrm{Co}_{27} \rightarrow {}^{60}\mathrm{Ni}_{28} + \mathrm{e}^- + \bar{\upsilon} + \gamma \tag{1.7}$$

Electron capture is a mode of radioactive decay in which the atomic nucleus absorbs an inner electron and changes a proton into a neutron and yields a neutrino

$$\rho + \mathrm{e}^- \rightarrow \eta + \upsilon \tag{1.8}$$

An example of decay by electron capture is iron-55, a common activation product with a half-life of 2.74 years (Audi *et al.*, 2003) in decommissioning wastes

$$^{55}\mathrm{Fe}_{26} + \mathrm{e}^- \rightarrow {}^{55}\mathrm{Mn}_{25} + \upsilon \tag{1.9}$$

Another mode of radioactive decay is isomeric transition in which the nucleus is in an excited state and emits gamma radiation with no changes in the number of protons or neutrons. An example of this the conversion of the medical isotope metastable technetium-99m to technium-99 with a half-life of 6.01 h (Unterweger *et al.*, 2016)

$$^{99\mathrm{m}}\mathrm{Tc}_{93} \rightarrow {}^{99}\mathrm{Tc}_{93} + \gamma \tag{1.10}$$

Another mechanism for radioactive decay is spontaneous fission. It is a characteristic of relatively heavy nuclei with an atomic number greater than about 90. The nucleus divides without external excitation into two fragments in addition to the release of neutrons and gamma radiation. An example of spontaneous fission is curium-250, an actinide that can occur in spent nuclear fuel. In addition to alpha and beta decay, spontaneous fission occurs about 74% of the time, yielding fragments of variable composition (Audi *et al.*, 2003).

1.4 Exposure and Dosimetry

1.4.1 *Dose*

The röntgen (R) is an early unit used to quantify exposure to ionizing radiation. It was named after Prof. Wihelm Röntgen, a German mechanical engineer and physicist. It is a measure of ionizing radiation in air generated by X-rays or γ radiation. One R produces 2.58×10^{-4} coulomb/kg of air (ATSDR, 1999). The röntgen is not a measure of the amount of radiation absorbed by solid matter, or the dose. Dose is the amount of radiation "deposited" on or in solid material such as human organs. Ultimately dose is a measure of the *risk* associated with the exposure to ionizing radiation.

A unit to quantify absorbed dose is the rad from "radiation absorbed dose" (Stabin, 2007). The rad is used predominantly in the US. It is defined as 100 ergs of ionizing energy absorbed by 1 g of matter. In the International System of Units (SI), absorbed dose is defined as the gray (Gy) which is equal to 1 J/kg of ionizing energy (Stabin, 2007). The gray was named after Dr. Louis Gray, an English physicist. One gray is equal to 100 rad.

1.4.2 *Equivalent Dose*

The quantification of absorbed dose as it impacts biological tissues is complicated. There are two primary impacts from exposure to ionizing radiation. Relatively large exposures greater than a threshold can result in tissue effects called "deterministic" because of the certainty of these effects occurring — and the effects are not cancers. The threshold dose is usually defined as the dose in which signs and symptoms of the effect on a specific organ or tissue can be detected. The time at which deterministic effects can be detected after exposure varies among tissues, which are classified as early-responding and late-responding. Relatively low-level exposures can result in a statistically increased chance of the occurrence of cancers and genetic effects. These effects are called "stochastic" because of the uncertainty of such effects occurring and have no threshold dose (IARC, 2000).

As discussed above, ionizing radiation can be composed of different type of particles with diverse kinetic energies. Moreover, ionizing radiation can have different effects on biological tissues depending on the type of radiation. Because the practice of health physics is most concerned with the protection of human beings, an approximation of dose is used to estimate absorbed dose and is called an equivalent dose. The equivalent dose is calculated by the use of weighting factors (w_R) that take into account the different types of radiation. These weighting factors are selected by the International Commission on Radiological Protection (www.icrp.org). The derivation of these factors is based in part on Linear Energy Transfer (LET) and Relative Biological Effectiveness (RBE). The LET is the rate at which a charged particle deposits and transmits kinetic energy as it moves through solid matter (IARC, 2000). LET is related to the "stopping power" of a material as LET divided by the density of the material. In general, the greater the LET of radiation, the more effective it is in producing biological damage (Cember and Johnson, 2008). The RBE is experimentally measured by exposing living cells such as cockroach embryos, rats and bacteria to radiation. For health physics applications, conservative RBE data are considered for protecting human health. The result are the weighting factors which are intended to "solve a practical problem" (Stabin, 2007). In other words, the weighting factors are used to protect people from routine exposures to radiation rather than to serve as a research tool for radiation-matter interactions. The weighting factors are considered as best estimates that are derived from the consensus of radiation safety regulators while working with industry interests and governmental directives. Note that the radiation weighting factor w_R was formally called the Quality Factor (Q) but it is not numerically equivalent in some cases. Moreover, weighting factors have changed with time as new toxicity data become available.

The equivalent dose (H_T) is a radiation-weighted dose. It is calculated by multiplying the absorbed dose by the weighting factor that is appropriate for the specific type of radiation. If the organ or tissue is exposed to only one type of radiation, then the equivalent dose is

$$H_T = D_T \times w_R \tag{1.11}$$

where H_T is the radiation-weighted dose and D_T is the absorbed dose of radiation.

The unit of H_T in SI units is (Sv) sievert which is equal to 1 J/kg. The unit sievert was named after Prof. Rolf Sievert, a Swedish medical physicist. In the US, the unit used is rem which is from röntgen equivalent in man. One Sv is equal to 100 rem. The weighting factor w_R has no units. To calculate the equivalent dose for a mixture of radiation types and energies exposed to different organs or tissues, a summation is taken of each

$$H_T = \Sigma w_R \times D_{T,R} \tag{1.12}$$

where H_T is the radiation-weighted or effective dose and $D_{T,R}$ is the absorbed dose of radiation averaged for each tissue.

1.4.3 *Effective Dose Equivalent*

The effective dose equivalent (H_E) is a tissue-weighted approximation to take into account the different susceptibilities of tissues and organs to ionizing radiation. The effective dose to each type of organ is multiplied by a dimensionless tissue weighting factor (W_T) (Cember and Johnson, 2008). This factor is chosen by the International Commission on Radiological Protection. As above, these weighting factors are used to protect people from routine exposures to radiation rather than to serve as a research tool for radiation-health studies. Given a multiplicity of organs, the effective dose equivalent is calculated by summing W_T multiplied by the effective dose (Eq. (1.12)) to each organ:

$$H_E = \Sigma W_T \times H_T \tag{1.13}$$

Example of Effective Dose

Suppose that a person held a class GTCC waste (see Chapter 4) in her lap. Her stomach absorbed 110 mSv equivalent dose and her bladder received 70 mSv. Using the tissue weighting factor for the stomach (0.12 from ICRP, 2007), and the bladder (0.04), the effective dose is

$$(110\,\text{mSv} \times 0.12) + (70\,\text{mSv} \times 0.04) = 16.0\,\text{mSv} \tag{1.14}$$

The summation of all the tissue weighting factors must equal to 1.0. In this example, only two organs were exposed. Therefore, for the sake of clarity, we can add to Eq. (1.14)

$$0\,\text{mSv} \times (1.0 - 0.16) = 0\,\text{mSv} \tag{1.15}$$

This addition of course does not change the effective dose calculation, but it illustrates that the effective dose (16.0 mSv) can be called the *whole-body effective dose.* The whole-body effective dose is the summation of the effective doses to each region of the human body (Stabin, 2007).

1.4.4 *Shallow- and Deep-Dose Equivalent*

The Shallow-Dose Equivalent is the external exposure dose equivalent to the skin or an extremity at a tissue depth of 70 μm averaged over an area of 1 cm.[2] The Deep-Dose Equivalent is the external whole-body dose exposure equivalent at a tissue depth of 1 cm (US NRC, 2019).

1.4.5 *Collective Effective Dose*

The collective dose (S) is the summation of all the effective doses received by a population of people. This type of dose can be used to evaluate the overall health effects of a process or accidental exposures to ionizing radiation by summing each effective dose per person times the number of people (N) in the population:

$$S = \Sigma H_{\mathrm{E}i} \times N_i \tag{1.16}$$

This type of dose may be more useful in medical applications of ionizing radiation rather than waste management.

1.5 Exposure and Biological Responses

The biological effects of ionizing radiation have been intensively investigated for decades (Cember and Johnson, 2008). The numerous observations and quantitative data have been derived from four diverse sources (IARC, 2000):

1. Occupational exposures such as uranium mining and milling, uranium fuel enrichment, nuclear reactor operation, nuclear fuel reprocessing, coal mining, industrial uses, military uses, radium-dial painters, research workers, and air crew exposures. This source also includes workers at the Nuclear Weapons Complex in the US (see Chapter 9), and the Russian Weapons Complex in the former USSR and the Russian Federation.
2. Patients exposed to medical radiation such as diagnoses, treatments, and medical experiments.
3. Survivors of the nuclear bombs dropped over Hiroshima and Nagasaki to help end World War II in Asia.
4. Exposures to nuclear weapons testing, and major accidents such as the Kyshtym accident in 1957 (USSR), the fire at the Windscale Piles in 1957 in the UK, and the accident at the Chernobyl Nuclear Power Plant in 1986 in the Ukraine.

The Committee on the Biological Effects of Ionizing Radiation of the National Academy of Science in the US (http://www.nasonline.org/) reviewed the information from the various sources and published reports called Biological Effects of Ionizing Radiation (BEIR). These reports have become influential compilations in establishing relationships between exposure to ionizing radiation and the standards to protect human health (Stabin, 2007). BEIR V (NRC, 1990), discusses health effects of relatively low levels of ionizing radiation. The authors of BEIR V supported the use of the Linear-No-Threshold (LNT) model for protection from ionizing radiation.

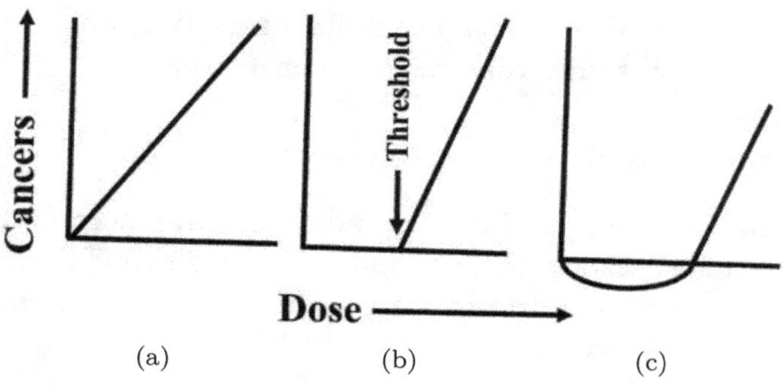

Figure 1.1. The LNT model (a), Threshold model (b), and the Hormesis model (c).

The LNT model is a simple dose-response curve (Fig. 1.1(a)) which assumes that an increase in dose uniformly results in an increase in a response — typically cancers such as leukemia, bone cancer, lung cancer, and thyroid cancer. The LNT model is essentially a linear regression that is extrapolated to the origin where information is lacking. The LNT model is radiologically conservative in that it implies that *any* exposure to ionizing radiation will pose some risk. In other words, there is *no* safe dose. The LNT model is controversial. The US Nuclear Regulatory Commission (US NRC) acknowledged there are not enough data available to conclusively establish a relationship between cancer and doses of ionizing radiation that are less than 100 mSv (10 rem) (US NRC, 2017). The US NRC and the US Environmental Protection Agency (US EPA) both advocate the use of the LNT model. The US NRC regards the LNT model as a regulatory tool for protecting human health from ionizing radiation.

Alternative dose-response relationships have been proposed. The Threshold Model implies that relatively low doses of ionizing radiation do not result in the development of cancers (Fig. 1.1(b)). The impacted cells recover and function normally. At a certain threshold, the incidence of cancers occurs and increases with an increase in dose (Stabin, 2007). Another concept is the Hormesis Model which advocates that relatively low levels of ionizing radiation

Table 1.1. Dose limits for occupational exposure based on the LNT model.

IAEA (2014). Occupational exposure limits for workers over the age of 18 years.

1. An effective dose limit of 20 mSv per year, averaged over five consecutive years.
2. An equivalent dose limit to the lens of the eye of 20 mSv per year, averaged over 5 consecutive years.
3. An equivalent dose of 500 mSv in a year to the extremities (hands and feet) or to skin.

US NRC (2019). Subpart C. Occupational dose limits for adults.

1. An annual dose limit of which of the following is the most limiting:
   a. An effective dose equivalent (or whole-body) limit of 5 rem (0.05 Sv) per year.
   b. The sum of the deep-dose equivalent plus the committed dose equivalent to any organ or tissue limited to 50 rem (0.5 Sv).
2. The annual limits to the lens of the eye, whole-body skin, and skin of extremities:
   a. A lens dose equivalent of 15 rem (0.15 Sv) per year
   b. A shallow-dose equivalent of 50 rem (0.5 Sv) per year to the whole-body skin or to the skin of any extremity.

are beneficial and reduce the incidence of cancer before negative impacts are possible at larger doses (Fig. 1.1(c)). The Committee on the Biological Effects of Ionizing Radiation concluded that the evidence for hormetic effects of ionizing radiation is inconclusive (NRC, 1990).

The LNT model currently prevails for the establishment of standards for protection from ionizing radiation. Table 1.1 provides a compilation of dose limits for occupational exposure that might be encountered while managing radioactive wastes. The International Atomic Energy Agency (IAEA, 2014) stated that "the IAEA safety standards are based on [the LNT hypothesis]. It is not proven — indeed it is probably not provable ... but it is considered the most radiologically defensible assumption on which to base safety standards."

1.6 Background Ionizing Radiation

The discussion about the LNT model may prompt the reader to ask, "What *are* relatively low-levels of ionizing radiation?" It may surprise the general public to learn that they are exposed to low levels of radiation every day. Radioactivity occurs naturally in the atmosphere, drinking water, soils, building stone, and in many foods. People are exposed to low levels of radiation from natural and artificial sources during their entire lifetimes.

The United Nations Scientific Committee on the Effects of Atomic Radiation estimated that the summation of all possible natural sources yields an average effective dose of ionizing radiation per person in the world is 2.40 mSv per year[1] (240 mrem/year) (UNSCEAR, 2008). When all potential artificial sources (mostly medical) of radiation are added to natural sources, the average effective dose per person is about 3.01 mSv/year (301 mrem/year) (Fig. 1.2). According to the National Council on Radiation Protection and Measure (in the US), the US population may be exposed to an average of 6.24 mSv/year (624 mrem/year) per person during their lifetime when both natural and artificial sources are combined (NCRP, 2009). In general, natural sources of ionizing radiation account for about half of the background radiation in the US, and medical applications currently account for about 48%.

The effective dose of natural radiation varies with altitude, geographical location, and lifestyle. For example, cosmic radiation (a collection of high-energy subatomic particles and ionized nuclei) will increase from about 0.26 mSv/year (26 mrem/year) at sea level to about 0.49 mSv/year (49 mrem/year) at an elevation of 1,000 m (NCRP, 2009). The earth's atmosphere provides shielding from cosmic radiation, but it interacts with the earth's upper atmosphere to produce a variety of interesting radionuclides such as ^{14}C, ^{10}Be, ^{26}Al, ^{36}Cl, ^{80}Kr, ^{32}Si, ^{22}Na, ^{35}S, ^{37}Ar, ^{33}P, and ^{39}Cl.

The largest single source of natural background radiation is inhalation of air containing gaseous radon (Fig. 1.2). The most stable

[1]The range of the mean was 1–13 mSv/year (UNSCEAR, 2008).

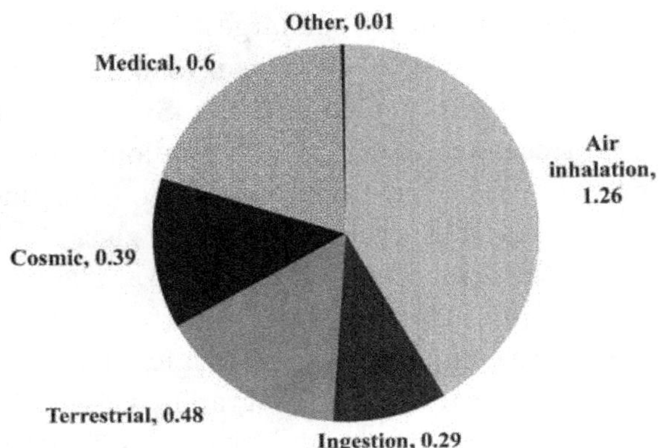

Figure 1.2. Global background ionizing radiation (in mSv/year). Air inhalation is dominated by radon. Ingestion refers to the consumption of foods and water containing K-40, C-14, U-238 and others. Terrestrial refers to radiation from soil, rocks, and building materials. Other refers to atmospheric testing of nuclear weapons, occupational exposure, the nuclear fuel cycle, and the Chernobyl accident (UNSCEAR, 2008). Reproduced by permission of UNSCEAR.

radionuclide (radon-222) is a source of alpha radiation and is a decay product from uranium (^{238}U) that is ubiquitous in soils and water. Thoron (^{220}Rn) is a decay product of thorium-232. Terrestrial sources in Fig. 1.2 refers to all the naturally occurring radionuclides in soils, rocks, surface water, and groundwater (see EPRI, 2014 for examples).

Ingestion in Fig. 1.2 refers to the dose of ionizing radiation within the human body from radionuclides derived from food, water, and air. Many foods contain radioactive isotopes of common elements. Any food or drink that contains potassium will also contain about 0.012% of the radionuclide potassium-40 which decays to Ca-40 by beta radiation with a half-life of 1.25×10^9 years (Audi *et al.*, 2003). Common foods such as potatoes, chicken and orange juice are all slightly radioactive because of ^{40}K. Milk contains about 74 Bq/L (2,000 pCi/L) of ^{40}K. The body of an adult human male contains about 3,700 Bq (100,000 pCi) of ^{40}K which yields a dose of about 0.18 mSv (18 mrem) to soft tissues each year (EPRI, 2014).

Increased applications of X-ray computed tomography, conventional radiography and fluoroscopy, and interventional fluoroscopy since the early 1990s account for the largest increase in the previous estimated effective dose of 0.36 mSv/year (360 mrem/year) in the US (NCRP, 2009). Nuclear medicine (0.77 mSv/year) uses ^{99m}Tc in the diagnosis and treatment of cancers, and radionuclides such as ^{125}I and ^{89}Sr in pellets for in-situ treatment of cancer. Consumer products (0.13 mSv/year) include older pottery, antique glass, bathroom tile, and jewelry in which uranium was used as a pigment (see also Chapter 3). Other products include cigarettes, smoke detectors, brick masonry, granite countertops, and lawn fertilizer. Occupational exposure includes medical staff, radiographers, dental hygienists, researchers, pharmacists, welders, and airplane and jet crews. This source also includes exposure during air travel; the dose depending on distance, altitude, and frequency of travel.

The LNT model implies that these background sources of ionizing radiation are potentially harmful. It is, however, impossible to avoid most of these natural and artificial sources. Protection from harm can only be achieved by minimizing exposure to *significant* doses — which is discussed in Section 1.7.

1.7 Minimizing Exposure to Ionizing Radiation

Reducing the risks of exposure to ionizing radiation is best introduced by the acronym ALARA as "as low as reasonably achievable" in the US and as low as reasonably practicable (ALARP) in the UK. The concept is intended to be the basis for practical protection from ionizing radiation in light of the implications of the LNT model of no safe dose.

When applied to the management of radioactive wastes, the ALARA/P principle can be expressed as three precautions: distance, stay time, and shielding. Distance means maximizing to the greatest extent possible the distance between the radioactive materials and the workers or the general public. The efficacy of distance is best illustrated by the Inverse Square Law: the intensity of radiation decreases in proportion of the reciprocal of the distance squared as

$$\mathrm{I_1 d_1^2 = I_2 d_2^2} \tag{1.17}$$

where I is the intensity of the ionizing radiation at locations 1 and 2, d is the distance 1 and 2 from the source of radiation.

Stay time means minimizing the amount of time that a person is exposed to ionizing radiation. Given that dose is equal to the dose rate x time, it follows that stay time is equal to a limit divided by the dose rate.

Example of distance in ALARA/P

If the effective dose rate 1 m from a waste container is 0.71 mSv/h (710 mrem/h) of gamma radiation, what is the effective dose rate at a distance of 20 m?

$$I_2 = (0.71\,\text{mSv/h})(1\,\text{m})^2/(20\,\text{m})^2 = 0.178\,\text{mSv/h} \tag{1.18}$$

Increasing the distance by 20 times, decreased the effective dose by two orders of magnitude.
Example of stay time in ALARA/P
How long can a worker operate in a 15-mSv/h (1.5-rem/h) field if we need to limit the dose to 1 mSv (100 mrem)?

$$\text{Stay time} = 1\,\text{mSv}/15\,\text{mSv/h} = 4\text{min} \tag{1.19}$$

The worker must leave the radiation field after *only* 4 min.

Shielding refers to the placement of physical barriers between radioactive sources and workers or the general public. An appreciation of how shielding is used to reduce exposure requires some understanding about how radiation reacts with solid matter.

1.8 Radiation Attenuation

The radiation from radioactive wastes is subject to scattering and absorption by solid matter. Because alpha radiation or particles are composed of divalent helium ions (He^{2+}), they interact significantly with solid matter resulting in relatively short travel paths. The particles lose kinetic energy during each interaction (Fig. 1.3). Eventually each alpha particle attracts 2 electrons and becomes

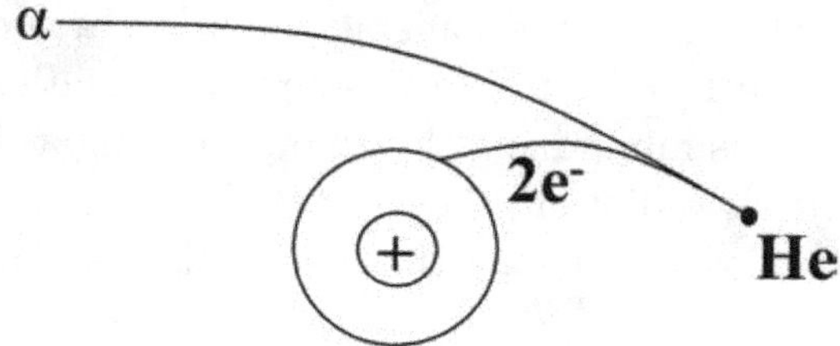

Figure 1.3. Attenuation of alpha radiation.

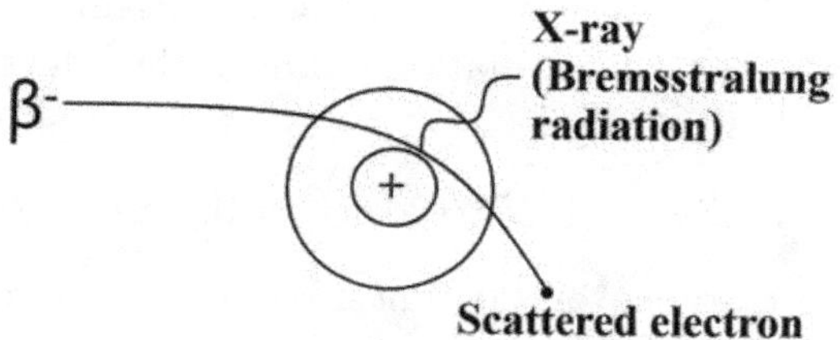

Figure 1.4. Attenuation of beta radiation.

helium gas at rest mass (i.e., no significant kinetic energy):

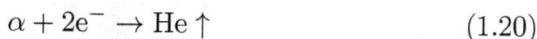

$$\alpha + 2e^- \rightarrow \text{He} \uparrow \tag{1.20}$$

Beta radiation can also ionize atoms in solid matter but to a lesser extent than alpha particles. As it moves near atoms, it can be slowed down in which some of the kinetic energy is covered into X-rays called Bremsstralung Radiation (Fig. 1.4). This is a form of secondary radiation resulting from the interaction of beta radiation with solid matter. As the beta particle loses its energy after collisions with atomic electrons, it eventually joins with orbital electrons. A positron will also experience atomic collisions, but will be attracted to electrons and the two will be annihilated releasing another secondary gamma radiation (http://www.radioactivity.eu.com).

Because gamma radiation is composed of relatively high-energy photons, they do not interact significantly with solid matter. Gamma radiation produces the photoelectric effect in which the kinetic energy of the gamma photon is transferred to an ejected electron, and the photon vanishes (Fig. 1.5). Compton scattering is another reaction

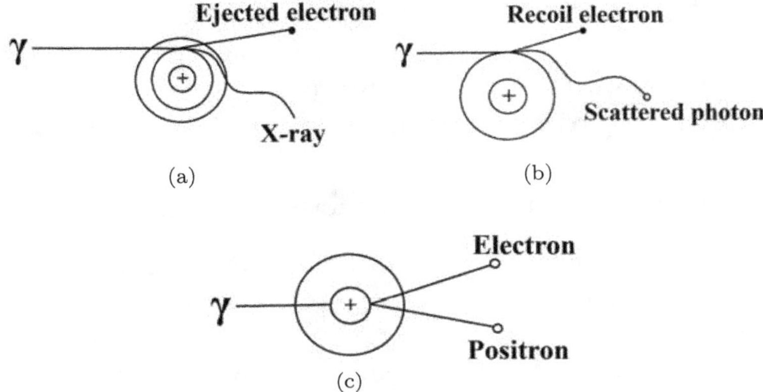

Figure 1.5. Attenuation of gamma radiation. (a) photoelectric effect, (b) Compton scattering, (c) pair production.

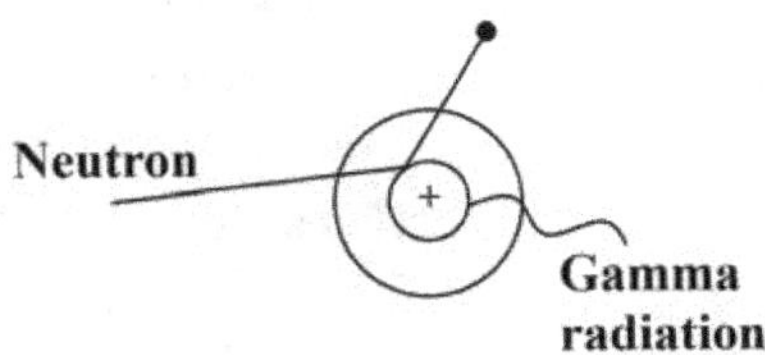

Figure 1.6. Elastic scattering of a neutron with secondary (gamma) radiation.

in which the kinetic energy of gamma radiation is dissipated between electron recoil and a scattered photon. During pair production, the gamma photons collide with the nucleus forming an electron–positron pair:

$$\gamma + \text{nucleus} \rightarrow \text{e}^- + \text{e}^+ \tag{1.21}$$

Because neutrons have no charge, they do not interact with the electrons of atoms. Neutrons interact with atomic nuclei by elastic scattering (deflected by collisions with the nuclei), and by absorption or capture in which they are combined with the nucleus. In elastic scattering, some of the kinetic energy is transferred to the nucleus leaving it in an excited stated. When the scattered nucleus relaxes, it emits gamma rays as a secondary radiation (Fig. 1.6). The greatest amount of kinetic energy can be transferred from the neutron to the nucleus when the latter has the same mass as the neutron. Therefore, the smaller the atomic mass of the shielding material, the more effective it is as a moderator. Water, plastic, and graphite are all used to slow neutrons by elastic scattering (US NRC, 2011).

1.9 Shielding Applied to Radioactive Wastes

Alpha particles are relatively large, and readily react with solid matter via Coulombic forces between the divalent charges and the electrons within the absorbent. Estimating the thickness of material needed for shielding can be accomplished by first experimentally measuring how far alpha particles travel in air as a function of their kinetic energy and the average distance they travel before coming to rest (referred to as the "range"). Their range in solid matter can be estimated using empirical relationships (see Cember and Johnson, 2008). The resulting solid-phase estimate is then divided by the density of the shielding material to calculate the thickness needed. For example, alpha particles from Pb-210 would require aluminum foil (with a density of 2.7 g/cm^3) with a thickness of only about 25 μm to effectively stop all the alpha particles (Cember and Johnson, 2008). The alpha radiation emitted from a radioactive waste or material is effectively shielded by sheets of paper. Given that paper would be impractical for a shipping package, alpha sources are typically managed in cardboard boxes for shielding from external radiation (Fig. 1.7).

Similarly, estimating the minimum thickness for shielding from beta radiation can be estimated experimentally by determining a relationship between the kinetic energy of the radiation and the range in different types absorbents such as air, water, or aluminum (Cember and Johnson, 2008). Therefore, given the maximum kinetic energy of the beta radiation, a minimum thickness of a shield can be calculated by dividing the range by the density of the material. For example, the aluminum foil above would need to be at least 0.93 cm thick to completely stop 2.27 MeV of beta radiation from Y-90 (Cember and Johnson, 2008). Radioactive wastes yielding beta radiation would be adequately shielded by containers made of wood, aluminum, or hard plastic (Fig. 1.7) (also see Chapter 8).

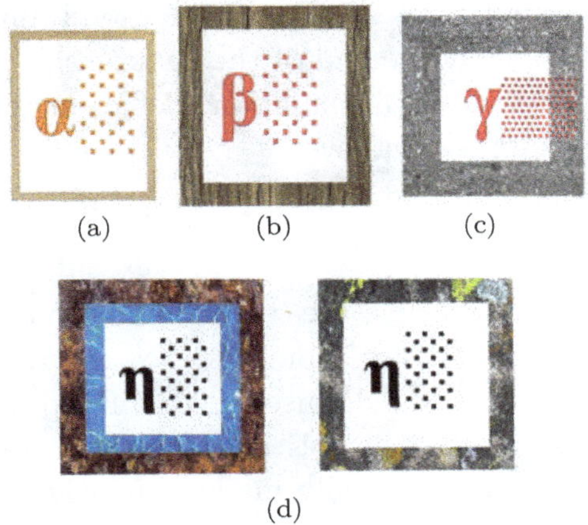

Figure 1.7. Representative materials used in shielding radioactive sources. (a) Alpha radiation can be stopped by a sheet of paper. For practical management, however, a carboard box will shield from external exposure. (b) Beta sources can be stopped by a wooden container (shown), sheets of aluminum, or rigid plastic. (c) Gamma radiation cannot be completely shielded. Relatively thick walls of lead, concrete, or steel are used to shield gamma radiation from wastes. (d) Neutron radiation can be shielded by combinations of iron, lead, and material composed of light elements such as hydrogen (left) or concrete mixed with elements that exhibit significant cross-sections for capturing neutrons (right).

Gamma radiation is more penetrating than alpha or beta radiation. The attenuation of gamma radiation can be reduced by increasing the thickness of shielding material, but it cannot be completely absorbed (Cember and Johnson, 2008). The reduction in the intensity of the gamma radiation can be calculated by Beer's Law (Saling and Fentiman, 2001),

$$I = I_0 \exp(-\mu_l t) \tag{1.22}$$

where I is the intensity of the shielded gamma radiation, I_0 is the intensity of the unshielded gamma radiation, μ_l is the linear attenuation coefficient, t is the thickness of the shielding material.

The linear attenuation coefficient is determined experimentally as the slope of a regression of transmitted gamma radiation versus the thickness of the absorbent which is the independent variable

(Saling and Fentiman, 2001). The reader will find some useful exercises on this topic at the end of the chapter.

Conceptually, the linear attenuation coefficient is the summation of three reactions discussed above:

$$\mu_{\mathrm{l}} = \mu_{\mathrm{pe}} + \mu_{\mathrm{cs}} + \mu_{\mathrm{pp}} \tag{1.23}$$

where the right-hand terms are the attenuation coefficients for the photoelectric effect (pe), Compton scattering (cs), and pair production (pp) (Cember and Johnson, 2008). For practical shielding, the linear attenuation coefficient is measured and used in designing waste containers that are typically made of relatively thick layers of concrete, steel, or lead (Fig. 1.7).

The reader may find the subject of neutron shielding an interesting topic beyond the scope of this chapter. It is an area of active international research that embraces neutron sources used in nuclear energy, industry and medicine (see Picha *et al.*, 2015 for example). Depending on the energy of the neutrons, shielding may be accomplished using iron or lead to first slow the neutrons, then absorb them. A layer of hydrogen-containing material can also be used to absorb the neutrons once they have been slowed by a thicker and more dense material (US NRC, 2011) (Fig. 1.7). Relatively small neutron sources are often shielded with polyethylene or paraffin. Shielding for larger neutron sources can be composed of concrete alone, or concrete made with elements that have a significant cross-section for capturing neutrons. These additives include boron carbide, boron oxide, sodium boron, iron–boron alloys, barite ($BaSO_4$), and others.

1.10 Review Questions

1. Define ionizing radiation. Why is it a concern to human health?
2. What is an effective half-life?
3. Describe alpha, beta, and gamma radioactive decay.
4. What is the difference between an equivalent dose and an effective dose equivalent of ionizing radiation?

5. A worker in a nuclear power plant was exposed to ionizing radiation from a shielded transfer cask containing used nuclear fuel. His lungs absorbed 0.17 mSv equivalent dose and his liver received 0.20 mSv. Using the tissue weighting factor for the lung (0.12 from ICRP, 2007), the liver (0.04), and the heart (0.12) what is the whole-body effective dose? How does the result compare with the occupational limit of an effective dose equivalent limit of 0.05 Sv per year?
6. Based on the measurements below, determine the half-life of an unknown isotope. What isotope might this be? Hint: calculate the slope of a regression of time versus the corrected radioactivity. The slope can be used to calculate the reaction constant k needed for solving half-life = 0.693/k.

Day	**Radioactivity**	**Background**
0	8,761	39
1	8,050	47
2	7,295	51
3	6,790	26
4	6,157	21
5	5,709	37
6	5,237	30
7	4,779	62
8	4,427	45
9	4,003	22
10	3,690	19

7. If we had a radioactive waste giving off 5 MeV of gamma radiation, how thick would lead or concrete need to be to reduce the radiation by 99%? Assume that the linear attenuation coefficient of lead is 0.482 and 0.067 cm^{-1} for concrete.
8. How much of a 0.5-Mev source of gamma radiation will be shielded by a 1.0-cm thick copper plate given that the mass attenuation coefficient is 0.0820 cm^2/g?

9. Describe the sources of information available that have been used to evaluate the biological impacts of ionizing radiation.
10. Describe the Linear-No-Threshold model for protection from ionizing radiation. What is the most significant implication of this model in managing radioactive wastes?
11. What is the average, annual effective dose of ionizing radiation per person in the world?
12. What is the largest single source of natural background radiation, and where does it come from?
13. Describe the attenuation of alpha, beta, and gamma radiation by solid matter. What is secondary radiation?
14. Given that radioactive wastes can emit alpha, beta, and gamma radiation *at the same time*, what type of material would be best to reduce occupational exposure to ionizing radiation?

Bibliography

Ashraf, M. A., Akib, S., Maah, M. J., Yusoff, I. and Balkhair, K. S. (2014). Cesium-137: Radio-Chemistry, Fate, and Transport, Remediation, and Future Concerns. *Critical Reviews in Environmental Science and Technology*, 44, pp. 1740–1793.

ATSDR. (1999). *Toxicological Profile for Ionizing Radiation.* (US) Agency for Toxic Substances and Disease Registry, Atlanta, Georgia.

ATSDR. (2004). *Toxicological Profile for Cesium.* (US) Agency for Toxic Substances and Disease Registry, Atlanta, Georgia.

Audi, G., Bersillon, O., Blachot, J. and Wapstra, A. H. (2003). The NUBASE Evaluation of Nuclear and Decay Properties. *Nuclear Physics A*, 729, pp. 3–128.

Cember, H. and Johnson, T. E. (2008). *Introduction to Health Physics*, 4th Ed., McGraw-Hill, Inc., New York, New York.

EPRI. (2014). *Assessment of Radioactive Elements in Coal Combustion Products.* Electric Power Research Institute. (Report number 3002003779), Palo Alto, California.

IAEA. (2014). *Radiation Protection and Safety of Radiation Sources: International Basic Safety Standards.* General Safety Requirements Part 3, International Atomic Energy Agency.

IARC. (2000). *IARC Monographs on the Evaluation of Carcinogenic Risks to Humans. Ionizing Radiation, Part 1: X- and Gamma (γ)-Radiation, and Neutrons*, v. 75, International Agency for Research on Cancer, Lyon, France.

ICRP. (2007). *The 2007 Recommendations of the International Commission on Radiological Protection.* Annals of the ICRP Publication 103, The International Commission on Radiological Protection.

NRC. (1990). *Health Effects of Exposure to Low Levels of Ionizing Radiation: BEIR V.* National Research Council, Washington, DC: The National Academies Press.

NCRP. (2009). *Ionizing Radiation Exposure of the Population of the United States.* National Council on Radiation Protection and Measurements. (Report 160), Bethesda, Maryland.

Picha, R., Channule, J., Khaweerat, S., Liamsuwan, T., Promping, P., Ratanatongchai, W., Silva, S., and Wonglee, S. (2015). Gamma and Neutron Attenuation Properties of Barite-Cement Mixture. *Journal of Physics: Conference Series*, 611, pp. 1–7.

Saling, J. H, and Fentiman, A. W. (2002). *Radioactive Waste Management.* Taylor & Francis, New York, New York.

Stabin, M. G. (2007). *Radiation Protection and Dosimetry.* Springer, New York, New York.

UNSCEAR. (2008). *Sources and Effects of Ionizing Radiation.* Volume I: Sources. United Nations Scientific Committee on the Effects of Atomic Radiation, New York, New York.

Unterweger, M. P., Hoppes, D. D., Schima, F. J., and Coursey, J. S. (2016). *Radionuclide Half-Life Measurements Data.* National Institute of Standards and Technology. Available at: https://dx.doi.org/10.18434/T41P42 [Accessed 13 January 2020].

US NRC. (2011). *Interactions of Neutrons with Matter.* Document 26. 0751 — H122 — Basic Health Physics, US Nuclear Regulatory Commission. Available at: https://www.nrc.gov/docs/ML1122/ML11229A619.html [Accessed 13 January 2020].

US NRC. (2017). Backgrounder on Biological Effects of Radiation. U.S. Nuclear Regulatory Commission. Available at: https://www.nrc.gov/reading-rm/doc-collections/fact-sheets/bio-effects-radiation.html [Accessed 13 January 2020].

US NRC. (2019). Subpart C — Occupational Dose Limits. US Nuclear Regulatory Commission. Available at: https://www.nrc.gov/reading-rm/doc-collections/cfr/part020/part020-1201.html [Accessed 13 January 2020].

Chapter 2

Radionuclides in Groundwater

"One thing, however, caused us great uneasiness–our water reserve was already half exhausted."

— Henry Lawson in *Journey to the Center of the Earth* by Jules Verne

2.1 Introduction

The mobilization of radionuclides from a disposal site begins with groundwater coming into contact with some type of waste package or container. If the structural integrity of the container has been compromised, radionuclides can leach from the waste. Waste containers can be damaged by overburden pressure, chemical corrosion of metallic barriers, or other long-term natural processes such as shearing forces resulting from bedrock movement.

The purpose of this chapter is to provide a concise treatment of the potential geochemical interactions of radionuclides with geomedia (rocks, minerals, soils, sediments), and their movement in groundwater after they have escaped from shallow land disposal or from a deep geological repository. It is not within the scope of this chapter to present a comprehensive discussion of environmental geochemistry or groundwater hydrogeology. Numerous textbooks such as Brookens (1984), Fetter (1993), and Wiles (2002) are available. A report by Siegel and Bryan (2003) also warrants serious study.

2.2 Radionuclides in Solution

Depending on the composition of the radioactive wastes, the radionuclides that could leach into groundwater could vary in type from activation products from low-level wastes, fission products from spent nuclear fuel, to uranium and its decay products from uranium milling operations. These radionuclides can have half-lives that range over several orders of magnitude, and result in the transformation of less stable radionuclides into more stable forms. It is, however, extremely convenient to recall that this potentially diverse and dynamic plethora of radioactive substances are all elements in the Periodic Table which will lead to a least a starting point in understanding their behavior in groundwater. The elements in the Periodic Table are arranged by common properties. For example, strontium is both a radioactive and a stable element in the Group 2 family with calcium. Like calcium, strontium will lose two electrons in solution, becoming a divalent ion (cation). Group 17 elements gain an electron in solution. For example, regardless of whether it is a stable or radioactive isotope, chlorine will gain an electron in solution becoming a negatively charged ion (anion). Radon is a decay product of radium, and belongs to Group 18, the Noble Gasses. Like argon and neon, radon is chemically inert. Transition elements can occur in multiple valences which influences their behavior in groundwater.

The pH of the groundwater is often the master variable that controls the specific form and concentration of radionuclides. The pH scale is the negative of the base-10 logarithm of the hydrogen ion activity. For example, the specific form of americium is pH-dependent. In acidic solutions, it occurs as a trivalent cation (Fig. 2.1). With a decrease in acidity, the hydroxide ion concentration of the solution increases and reacts with americium:

$$Am^{3+} + OH^- \rightleftharpoons AmOH^{2+} \tag{2.1}$$

A subsequent increase in pH results in a greater amount of hydroxide ions, and an increase in americium hydroxide forms such as $Am(OH)_2^+$.

Another important variable that is not as familiar as pH is the oxidation–reduction potential or redox potential (Eh). Oxidation is the donation of electrons whereas reduction is the acceptance of electrons from ions and molecules. Oxygen is the major electron acceptor in solution, and chemically reduced groundwater is oxygen

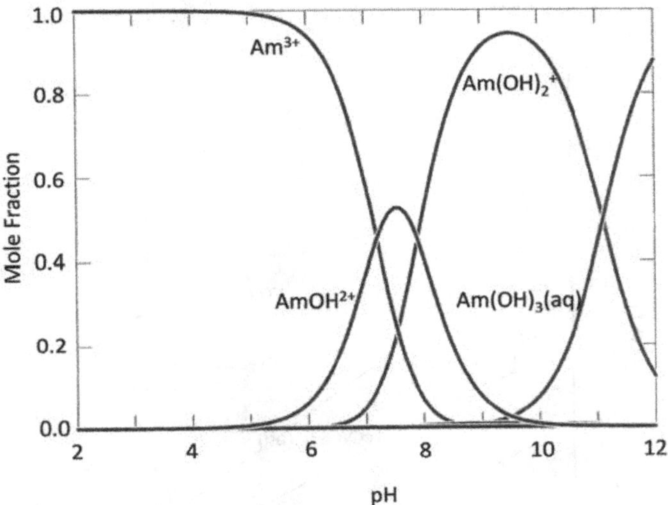

Figure 2.1. Distribution of americium species at a concentration of 10^{-12} M (830 pCi/L) in deionized water at 298.15 K, 0.1 MPa (Cantrell and Felmy, 2012).

poor. The chemical stability of water is defined by an upper boundary in which water is oxidized to oxygen gas. The lower boundary of water stability is defined by its reduction to hydrogen gas. These boundaries are best illustrated by a Pourbaix diagram (Fig. 2.2) named after the Belgium chemist Marcel Pourbaix. They are more commonly called Eh–pH diagrams. Such diagrams are useful for understanding the dominant thermodynamic forms of radionuclides in solution, and for qualitative predictions about their behavior while migrating away from a disposal site. For example, the valence of technetium depends on the Eh of the solution. Under oxidizing conditions, it occurs in combination with oxygen as Tc (VII) in the form of an anion (TcO_4^-). Under reduced conditions, the dominant form of technetium is TcO_2 (Fig. 2.2) in which technetium has been reduced to Tc (IV). The change in oxidation state has a profound impact on the mobility of this fission product in groundwater. The valence of some radionuclides, however, is not sensitive to changes in Eh. For example, cesium occurs as Cs^+ throughout the entire stability field of water.

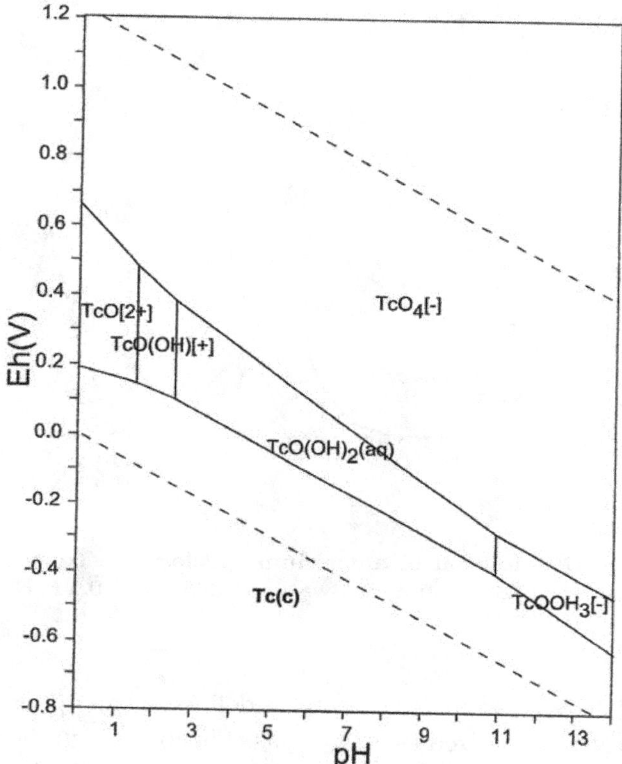

Figure 2.2. A Pourbaix or Eh–pH diagram for technetium. The upper dashed line is oxidation of water into oxygen gas. The lower dashed line is the reduction of water into hydrogen gas. These two dashed lines define the stability of water. Under oxidizing conditions, the most stable thermodynamic form of technetium in solution is TcO_4^-. Under reducing conditions, the most stable form of technetium is technetium hydroxide as a solid (s) phase (from GSJ, 2005). Used with permission from the Geological Survey of Japan.

The charge of the ion will determine how it reacts with other ions in solution, and how it will react with geomedia as it moves in groundwater away from the disposal site. Regardless of the type, chemical reactions are often written as a cation (A^+) reacting with an anion (B^-) as reactants yield AB as the product:

$$A^+ + B^- \rightleftharpoons AB \tag{2.2}$$

The reaction of A^+ with B^- could be the result of the sharing of electrons yielding a covalent bond, or as a result of electrostatic attractions yielding an ionic bond. At chemical equilibrium, the rate of the forward reaction yielding AB is the same as the reverse reaction yielding A^+ and B^-, and the concentration of the product is constant with time. Some chemical reactions in water equilibrate within seconds while others such as reactions of water with silicate minerals may require years. At equilibrium, it is convenient to define a given reaction by its equilibrium constant (K_{eq}) which is calculated as the ratio of the activities of the products to the reactants:

$$K_{\text{eq}} = \frac{\text{AB}}{(\text{A}^+)(\text{B}^-)} \tag{2.3}$$

Activity is defined as the solution concentration [C] corrected by an activity coefficient (γ) to correct for non-ideal behavior in non-dilute solutions: $(\text{C}_i) = \gamma \text{C}_i$

2.3 Reactions between Radionuclides and Groundwater

Whether groundwater is a shallow, chemically dilute, drinking water source, or a deep, concentrated, saline solution, it will contain various naturally occurring ions and dissolved constituents that will react with radionuclides that have escaped from the disposal area. Regardless of type and depth, groundwater will typically contain six major constituents: calcium, magnesium, and sodium as cations, and bicarbonate (HCO_3), sulfate (SO_4) and chloride as anions. Depending on the types of rocks in the subsurface, secondary constituents include iron, strontium, potassium, carbonate (CO_3^{2-}), nitrate (NO_3^-) and fluoride (F^-). Moreover, dissolved gases such as sulfide, methane, carbon dioxide, and oxygen may also be present.

For example, the dominant form of americium at pH less than about 6 is a trivalent cation (Am^{3+}). In a solution containing sulfate and carbonate at a pH greater than 6, other reactions will occur (Table 2.1).

Table 2.1. Americium speciation reactions (Cantrell and Felmy, 2012).

Reaction	Log K at 298.15 K	Reaction number
$Am^{3+} + 2H_2O \rightleftharpoons Am(OH)_2^+ + 2H^+$	−15.0	2.4
$Am^{3+} + 3H_2O \rightleftharpoons Am(OH)_{3(aq)} + 3H^+$	−26.2	2.5
$Am^{3+} + SO_4^{2-} \rightleftharpoons AmSO_4^+$	3.30	2.6
$Am^{3+} + 2HCO_3^- \rightleftharpoons Am(CO_3)_2^- + 2H^+$	−7.758	2.7

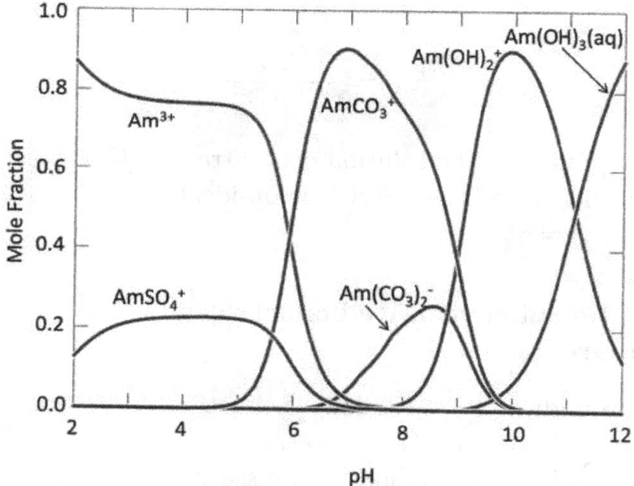

Figure 2.3. Distribution of americium species at a concentration of 10^{-12} M (830 pCi/L) in water containing sulfate and carbonate ions at 298.15 K, 0.1 MPa (adapted from Cantrell and Felmy, 2012).

Reactions (2.4) and (2.5) are hydrolysis reactions in which americium reacts with water, forming americium hydroxide. Reaction (2.7) converts the trivalent cation into a carbonate anion. The relative distribution of the various americium species depends on the pH of the solution (Fig. 2.3). The americium sulfate ion occurs under acidic conditions whereas the americium carbonate ions dominate in near-neutral pH conditions. Hydroxide forms of americium are dominant at pH greater than about 9.

2.4 Precipitation and Dissolution Reactions

Radionuclides in groundwater can precipitate as solid phases when they react with other ions. Solubility can be defined as the amount of a given mineral that can dissolve into a solution. When a chemical precipitate begins to form, the solubility of the mineral has been exceeded, and the solution is said to be supersaturated with respect to that mineral phase. During the dissolution process, the solution is said to be undersaturated with respect to that mineral phase. The relative degree of saturation is often evaluated by a saturation index (Ω) which can be defined as

$$\Omega = \log \frac{\text{IAP}}{K_{\text{eq}(t)}} \tag{2.4}$$

where IAP is the ion activity product, $K_{\text{eq}(t)}$ is the equilibrium constant at temperature, t.

For example, the solubility of strontium-90 can be treated as the solubility of stable strontium-88. When reacted with different anions, the solubility of the solid phases can be ranked as $SrCO_3$ (most insoluble) > $SrSO_4$ > SrF_2. Strontium carbonate (strontianite) is the most likely solubility-limiting solid phase in a solution containing carbonate ions. The reaction becomes

$$Sr^{2+} + CO_3^{2-} \rightleftharpoons SrCO_3 \downarrow \tag{2.5}$$

The equilibrium constant for reaction 2.5 is 5.3661×10^{-10} at 298 K (25°C) (Busenberg *et al.*, 1984). As an example, assuming that the concentration of Sr^{2+} is 3.2 mg/L and CO_3^{2-} is 15.3 mg/L, will strontium carbonate precipitate?

$$3.2\ \text{mg Sr/L} = 3.6521 \times 10^{-5}\ \text{moles/L and}$$
$$15.3\ \text{mg}\ CO_3\text{/L} = 2.5496 \times 10^{-4}\ \text{moles/L}$$

The ion activity product for this reaction is (Sr^{2+}) (CO_3^{2-}). If for the sake of simplicity concentrations are assumed to equal activities which have no units, then

$$\Omega = \log \frac{(3.6521 \times 10^{-5})(2.5496 \times 10^{-4})}{(5.3661 \times 10^{-10})}$$

$$\Omega = 1.24$$

Because Ω is greater than 0, strontium carbonate is thermodynamically poised to precipitate from solution at equilibrium. In general, when $\Omega = 0$, the solution is in equilibrium with respect to the solid phase. When Ω is less than 0, the solid phase should dissolve if it is present. When Ω is greater than 0, the solid phase should precipitate at a rate controlled by the kinetics of the reaction.

The mobility of some types of radionuclides in groundwater may be limited by the formation of sparingly soluble carbonate or hydroxide solid phases, whereas some are not. For example, iodine-129 occurs as a monovalent ion ($^{-}$I) and reacts with potassium and sodium. However, like most halides, both solid-phase potassium and sodium iodide are too soluble in solution to form under typical groundwater conditions.

2.5 Sorption and Desorption Reactions

The movement of ionic radionuclides in groundwater may be retarded by reacting with charged surfaces associated with geological media in subsurface environments. The types of geomedia that have the ability to remove or sorb radionuclides vary from minerals, rocks (mixtures of minerals), consolidated and unconsolidated sediments, and soils formed from the weathering of a parent material. The origins of these charged surfaces can be generalized as those resulting from a pH-dependent, variable charge, and/or a stronger permanent charge derived from atomic substitutions. As an example of the former, quartz (SiO_2) is the most common mineral in the earth's crust, and a common component of aluminosilicate minerals which form many rocks. Quartz can sorb cesium from solution, and the extent of sorption is greatest in alkaline solutions (Cornell, 1993). The surface of quartz (shown by the symbol $\equiv$) can be thought of hydroxyls that can accept or donate a proton which determines surface charge.

$$\equiv\text{SiOH} \rightleftharpoons \equiv\text{SiO}^- + \text{H}^+ \tag{2.6}$$

At pH greater than about 2, the surface of quartz has a net negative or anionic charge that attracts cations like Cs^+. Depending on the pH of the groundwater, some iron, manganese, and titanium oxides can also possess a variable charge, and can contribute to the sorption of cationic radionuclides.

The efficacy of this pH-dependent charge can be dwarfed, however, by the influence of geomedia having a permanent surface charge. The permanent charge is the result of isomorphic substitutions of ions that take place during crystallization. Within silicate minerals, Al^{3+} can replace Si^{4+} that is in tetrahedral coordination with oxygen, and Mg^{2+}, Fe^{2+}, and Fe^{3+} can substitute for Al^{3+} that is in octahedral coordination with oxygen. The size of the ions, rather than the charge, controls the extent of substitutions. These substitutions will result in an excess negative charge on particle surfaces which are balanced by the presence of cations. In solution, these cations can be sorbed from solution, and can be readily exchanged with other ions such as cationic radionuclides:

$$\equiv\text{Ca} + \text{Sr}^{2+} \rightleftharpoons \equiv\text{Sr} + \text{Ca}^{2+} \tag{2.7}$$

This reaction is called ion exchange, and the ability of the sorbent to remove and retain cations is called the cation exchange capacity (CEC). The space occupied by calcium in reaction (2.7) is often referred to as an exchange site.

Desorption is the reverse reaction in Eq. (2.7) and can be induced by changes in groundwater composition. For example, if the source of the radionuclide is exhausted, then the concentration in the liquid phase would decrease. Such a decrease would create a concentration gradient away from the sorbed phase, and the radionuclide would be released into solution. When compared to the number of studies in which the extent of radionuclide sorption was measured under laboratory conditions, desorption has not received as much attention. It has been observed, however, that the sorption of radionuclides may not be fully reversible. That is, once sorbed, only a fraction of the sorbate can be induced back into solution under environmentally relevant conditions. This lack of sorption reversibility is called hysteresis which means the reverse reaction path does not match the forward reaction path. It has been proposed that the lack of reversibility may be the result of sorbed species diffusing into the sorbent matrix, making it less accessible to an extracting fluid. Also, the formation of post-sorption chemical bonds may contribute to hysteresis.

Sorption and desorption are typically measured under laboratory conditions by equilibrating solutions with known concentrations of a radionuclide with disaggregated samples of geomedia. In practice, a constant mass of sorbent can be equilibrated with different concentrations of the radionuclide, or different masses of sorbent are equilibrated with solutions containing the same concentration of the radionuclide. Roy *et al.* (1992) provided detailed procedures for measuring sorption using either approach, in addition to background information about sorption mechanisms, sorbent preparation, and data interpretation.

After the attainment of equilibria, the concentration of the radionuclide remaining in solution is measured, and the mass of sorbate per mass of the sorbent is calculated by difference as

$$\frac{(C_0 - C)V}{m} = \frac{x}{m} \tag{2.8}$$

where C_0 is the initial concentration of the radionuclide in solution, C is the concentration of the radionuclide after contact with the sorbent, V is the volume of the solution containing the radionuclide, m is the mass of the sorbent and x is the mass of the sorbate.

Desorption can be measured experimentally by removing a known volume from the equilibrated solution and replacing it with the same volume of solution that does not contain the radionuclide under study. The diluted suspension is then equilibrated again, and the amount of sorbate can be calculated by a mass balance.

$$\frac{(C_0V - C_1V_r - C_2V)}{m} = \frac{x}{m} \tag{2.9}$$

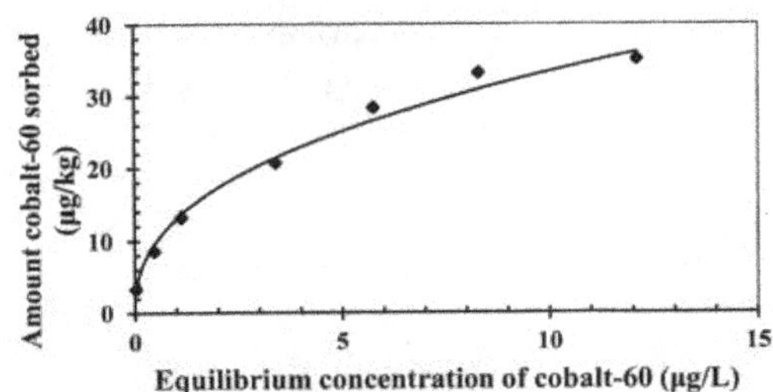

Figure 2.4. Sorption isotherm of cobalt-60 by illite at pH 6 and 298 K. The solid curve is a statistical regression through the experimental data points yielding a Freundlich Equation of $(x/m) = 13.0C^{0.41}$ with correlation coefficient of $r^2 = 0.994$.

where C_1 is the concentration from the sorption measurement, C_2 is the concentration of the radionuclide after desorption, and V_r is the volume of solution removed to induce desorption.

In practice, 3–5 desorption measurements may be needed to describe the reverse reaction. Sorption data are typically analyzed by plotting solution concentrations on the x-axis and the amount sorbed (x/m) on the y-axis (Fig. 2.4). The outcome is called a sorption isotherm that graphically illustrates the relationship between equilibrium concentrations and the amount sorbed at a constant temperature. Sorption data are typically generalized by the application of a statistical regression because of experimental variation. An often-used mathematical relationship is a power function

$$y = ax^b \tag{2.10}$$

which is the mathematical equivalent to the Freundlich Equation, named after the German chemist, Herbert M. F. Freundlich

$$\frac{x}{m} = K_f C^{1/n} \tag{2.11}$$

where x/m is the mass of sorbate divided by the mass of sorbent, K_f is the Freundlich constant, $1/n$ is the Freundlich exponent, and C is the equilibrium concentration of the radionuclide.

The Freundlich equation was originally used to model the adsorption of gases by solid surfaces, but it was borrowed for use in solid–liquid systems.

When the isotherm is linear, that is when 1/n is equal to 1.0, the Freundlich equation is often re-written as

$$\frac{x}{m} = K_d C \tag{2.12}$$

where K_d is referred to as a linear sorption constant or a solid–liquid partition coefficient. K_d values are essential for conducting transport modeling for a site assessment. They must be, however, interpreted with caution because the magnitude of these values can depend on the experimental conditions under which they were measured such as pH, redox conditions, time, and the chemical composition of the liquid phase such as deionized water as compared with a deep-saline brine.

Case study: Bentonite

Bentonite is being considered as a clay barrier in proposed geological repositories in various countries. The principal reactive component in bentonite is an expandable clay mineral called montmorillonite that has a significant capacity to sorb cesium by ion exchange (Cornell, 1993). Studies such as Missana *et al.* (2004) have concluded that cesium sorption by bentonite occurs in two steps: a rapid exchange reaction followed by a slower diffusion-driven reaction that results in sorption hysteresis.

2.6 Hydrogeology of Groundwater

For this chapter, groundwater will be regarded as the water below the water-unsaturated zone that flows in response to pressure gradients or changes in elevation (gravity). Groundwater may be the below-ground component of the Hydrologic Cycle that is subject to recharge by precipitation, or the byproduct of former climatic or geological conditions such as meltwaters from Pleistocene glaciers, or deeper waters derived from ancient marine environments that have not participated in the Hydrologic Cycle in millions of years.

2.6.1 *Hydraulic Conductivity*

Groundwater flows through porous media at a rate that depends on the driving force — the gradient — and the ability of the material to transmit water (the water-saturated hydraulic conductivity). Henri Darcy, a French engineer, proposed that for a volume of homogeneous, isotropic material that

$$\frac{K_{\text{sat}} A \Delta h}{L} = Q \tag{2.13}$$

where K_{sat} is the water-saturated hydraulic conductivity, A is the cross-sectional area of the volume of porous material, Δh is the change in elevation or hydraulic gradient (i), L is the distance of the flow path, and Q is the flow rate.

Darcy's law (Eq. (2.13)) can also be illustrated by Fig. 2.5.

Hydraulic conductivity emerges as another essential property that must be measured or estimated for conducting a site assessment for a disposal site. It can range over 10+ orders of magnitude. For example, a relatively homogeneous layer of sand-sized particles will

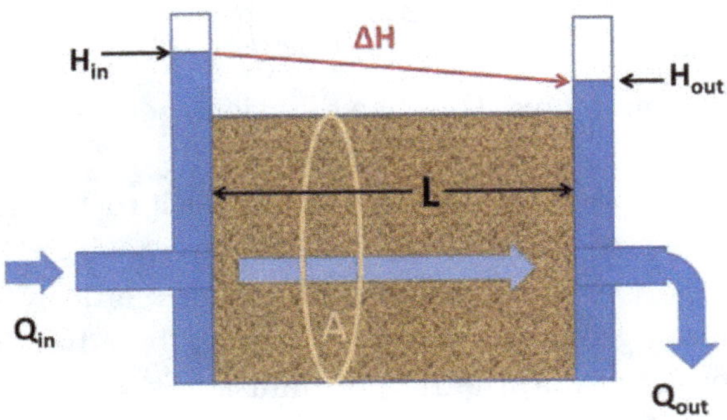

Figure 2.5. **Graphical illustration of Eq. (2.13). The difference in elevation ($H_{\text{in}} - H_{\text{out}}$) results in flow ($Q_{\text{out}}$) through cross-section A along flow path L.**

readily transmit water with a K_{sat} of 10^{-3} to $1\,\text{cm/s}$. The K_{sat} of soils and sediments can range from about 10^{-3} to $10^{-6}\,\text{cm/s}$. K_{sat} values less than $10^{-7}\,\text{cm/s}$ are generally regarded as relatively impermeable on a field scale, and are characteristic of compacted clays, and unfractured rocks such as limestone and granite. Fractured rocks, however, can be more conductive, depending on the width and length of the fractures, and the degree of continuity of the fractures.

Hydraulic conductivity can be measured under laboratory conditions using soil and rock core samples and under field conditions by a variety of methods. These field methods vary from adding or removing a known volume of water at the surface to measure infiltration, to more elaborate methods in which well-pumping tests are applied to measure changes in groundwater flow and elevation when a pump is in operation.

2.6.2 *Transport by Diffusion*

Radionuclides can still move in relatively impermeable materials. On a molecular level, no type of geomedia is completely impermeable. Radionuclides will spontaneously move from an area of greater concentrations toward an area of lesser concentrations by a process called diffusion. This type of movement is described by Fick's first law that was named after the German physiologist Adolf E. Fick

$$F = -D_l \left(\frac{dC}{dx} \right) \tag{2.14}$$

where F is the mass flux, D_l is the diffusion coefficient along path l, and dC/dx is the 1D concentration gradient.

Diffusion can occur within solid phases such rocks and clays, and in groundwater even if the liquid phase is not moving. Diffusion has the potential to transport radionuclides from relatively inaccessible solid phases into groundwater that is moving by advection (Section 2.6.3). In porous media, the rate of diffusion is slower than that in water because of the longer paths water must take around particles. Therefore, in porous geomedia, an effective diffusion coefficient can be defined as

$$D^* = \omega D_w \tag{2.15}$$

where D^* is the effective diffusion coefficient, ω is the related to tortuosity, another term experimentally measured under laboratory conditions, and D_w is the diffusion coefficient in water.

2.6.3 *Transport by Advection*

The transport of radionuclides by flowing groundwater is called advective transport. In order to illustrate an application of sorption constants (Eq. (2.12)), the term retardation (R) is defined as

$$R = \frac{\text{velocity of groundwater}}{\text{velocity of the radionuclide}} \tag{2.16}$$

$$= 1 + \frac{\rho_b K_d}{\theta} \tag{2.17}$$

where ρ_b is the dry bulk density, θ is the volumetric water content, and K_d is the sorption constant.

The influence of retardation of a radionuclide under advective flow with no dispersion can be calculated as

$$x = \frac{tK_{\text{sat}}i}{R\eta_e} \tag{2.18}$$

where x is the distance traveled, t is the time, K_{sat} is the saturated hydraulic conductivity, i is the hydraulic gradient, R is the retardation term from Eq. (2.16), and η_e = effective porosity.

The porosity (η) of porous media is defined as

$$\eta = \frac{V_v}{V_0} \tag{2.19}$$

where V_v is the volume of the void spaces or pores between the particles that comprise the matrix of the geomedia, V_0 is the total volume.

The effective porosity in Eq. (2.18) is the fraction of the total porosity that can conduct a fluid. Effective porosity does not include discontinuous and dead-end pores.

Case study

Diffusion coefficients can be measured under laboratory conditions or estimated by field observations. For example, Lu *et al.* (2008) created a laboratory-scale diffusion cell that they used to measure the diffusion of iodide-125 through sections of granite that was collected at a proposed geological repository in China. They reported an effective diffusion coefficient for ^{125}I-in the granite of 2.44 to $2.72 \times 10^{-12}\,m^2/s$.

2.6.4 *Transport by Dispersion*

Dispersion is a process in which radionuclides spontaneously spread out in porous media in flowing groundwater. At the microscopic level, dispersion results from the flow through different pore sizes and flow paths around particles of different sizes and shapes. At larger scales, dispersion results from geological heterogeneities and anisotropic materials. The overall effect of dispersion is that the concentrations of radionuclides moving away from a disposal site are diluted by mixing with groundwater.

Because dispersion results from a number of mechanisms, it is difficult to rigorously quantify its impact on the movement of radionuclides. For practical applications, a dispersion coefficient can be calculated from Lyman *et al.* (1992) as

$$D = \alpha_l V + D^* \tag{2.20}$$

where D is the dispersion coefficient, α_l is the longitudinal dispersivity of the media, V is the average linear groundwater velocity, and D^* is the effective diffusion coefficient from Eq. (2.15).

The longitudinal dispersivity component is best estimated from laboratory measurements of flow in soil columns using a non-reactive tracer.

2.6.5 *Transport by Advection–Dispersion–Reaction*

Radionuclide transport under the combined processes of advection, dispersion, and retardation in a water-saturated, homogeneous, isotropic materials under steady-state flow conditions along flow path x can be written in a 1D form (Freeze and Cherry, 1979) as

$$\frac{\partial C}{\partial t} = D_x\left(\frac{\partial C^2}{\partial x^2}\right) - V_X\left(\frac{\partial C}{\partial x}\right) - \left(\frac{\rho_b}{\theta}\right)\left(\frac{\partial S}{\partial t}\right) \tag{2.21}$$

where C is the concentration of the radionuclide in solution, D_x is the effective dispersion coefficient along flow path x, V_x is the mean convective flow along flow path x, ρ_b is the bulk density of the geomedia, θ is the volumetric water content, S is the amount of sorbate per mass of sorbent (x/m in Eq. (2.11)), and t is the time.

Equation 2.21 can be rearranged as

$$R\left(\frac{\partial C}{\partial t}\right) = D_x\left(\frac{\partial C^2}{\partial C^2}\right) - V_x\left(\frac{\partial C}{\partial x}\right) \tag{2.22}$$

where R is the retardation factor from Eq. (2.16).

The analytical solution of this second-order differential equation (Ogata, 1970) is

$$\frac{C}{C_0} = 0.5\left\{\operatorname{erfc}\left(\frac{(x - Vt^*)}{2(D_x t^*)^{0.5}}\right) + \exp\left(\frac{V_x}{D_x}\right)\operatorname{erfc}\left(\frac{(x + Vt^*)}{2(D_x t^*)^{0.5}}\right)\right\} \tag{2.23}$$

where C/C_o is the ratio of the radionuclide concentration in groundwater at time t and at distance x from the initial concentration C_0, erfc is the complimentary error function, V is the average linear pore-water velocity, D_x is the dispersion coefficient, T^* is the retarded time (actual time divided by the retardation factor R), and x is the distance of migration.

The reader should gain a better insight about the application of Eq. (2.23) by solving problem 6 and the end of this chapter.

2.6.6 *Transport by Stochastic Models*

A stochastic approach to estimate hydrogeological parameters relevant to radionuclide transport is used when the geomedia under study is neither homogeneous nor isotropic. A stochastic approach describes the probabilistic distribution and uncertainty of various characteristics such as porosity, hydraulic conductivity, or solute transport

(Fetter, 1993). A stochastic approach is site-specific, and the resulting model constructed and tested by iteration as additional data are collected. For example, the Pre-Cambrian bedrock at Forsmark, Sweden, is characterized as having extensive open fractures to a depth of about 200 m (Stephens *et al.*, 2015). This site was chosen for a geological repository for spent nuclear fuel (see Chapter 6). Site assessments were conducted by collecting field data at lineaments (linear topographic depressions that suggest the presence of a fault or fracture), and at fractures in areas having exposed bedrock surfaces, or in tunnel walls in excavations specifically created for the site assessment. Statistical models were then used to generalize the measured data, then to predict the likely properties of fractures and other characteristics in the bedrock in areas where data collection was not possible. The reader is directed to Follin (2008), for details about the site assessment and modeling at Forsmark.

2.7 Groundwater Monitoring

Groundwater monitoring may be required by regulatory agencies during the construction, operation, and the post-closure period of a shallow or deep repository for radioactive wastes. The overall purpose of monitoring is to detect adverse impacts on groundwater quality resulting from some type of failure at the disposal site. Both the design and scope of a monitoring program will be site specific. There are, however, two general tasks that would be endemic to any program:

1. Groundwater is often not directly accessible. The composition and movement of groundwater can only be *inferred* by collecting water samples from monitoring wells that are dedicated to a monitoring program. The number, depths, relative positions, and sampling protocols used for the monitoring wells, and the selection of solutes to measure will be site-specific and should be based on the results of detailed site-assessment studies.

2. The chemical composition of the groundwater samples must be documented. Moreover, the natural variation in composition must be assessed such as changes in response to seasonal variations, or long-term climatic changes. Depending on the size and depth of the proposed disposal site, it may take years to confidently assess the long-term variability in the chemical composition of groundwater.

2.8 Review Questions

1. At pH 8 and an Eh of 200 mV, what are the most thermodynamically stable forms in solution of the activation products sodium-24 and molybdenum-93?
2. A sorption coefficient of 3.14 mL/g has been measured for the sorption of radium-226 by samples of a subsurface clay layer. Given the following measured or estimated properties for this clay layer, how far would ^{226}Ra move in 20 years if we neglect dispersion?

Bulk density (ρ_b)	$= 1.69\,\text{g/cm}^3$
Volumetric water content (θ)	$= 0.2\,\text{cm}^3/\text{cm}^3$
Hydraulic conductivity (K_{sat})	$= 2.1 \times 10^{-5}\,\text{cm/s}$
Hydraulic gradient (i)	$= 0.01\,\text{cm/cm}$
Effective porosity (η_e)	$= 0.13\,\text{cm}^3/\text{cm}^3$

3. In question 2, assume that in the future that the subsurface clay is being exposed to an increase in the hydraulic gradient by 1,000%. This increase could be the result of the land surface being flooded by glacial meltwaters, or because the surface has subsided below the ocean because of long-term tectonic activity or because of an increase in sea level brought about by climate change. What effect would this event have on the estimated distance of transport of ^{226}Ra?

4. If the effective diffusion coefficient for $^{125}I^{-}$ in unfractured granite is $2.44 \times 10^{-12}\,m^2/s$, how far will it move in 1,000 years? Go to the Molecular Diffusion Calculator at http://www.earthwardconsulting.com/diffusion_calculator.htm

   Assume that input C/C_o $= 1 \times 10^{-3}$
   Porosity (φ) $= 0.005$
   Tortuosity (τ) $= 0.003$

5. Given sorption constants (K_d, cm^3/g) in the table below, how would you rank the relative mobilities of these radionuiclides in groundwater if they were released from a disposal site?

Radionuclide	**$K_d(cm^3/g)$**
^{226}Ra	70
^{232}Th	57,000
^{240}Pu	1,800
^{90}Sr	15

6. Assume that there is a continuous flow of groundwater downward from a geological repository. The concentration of iodide-129 in the groundwater as it leaves the repository is 210 mg/L and does not change with time. The iodide-129 is moving downward 50 m through an argillite layer that is homogeneous and isotropic. After 100 years, what will be the concentration of iodide-129 when it reaches the bottom of the argillite layer?

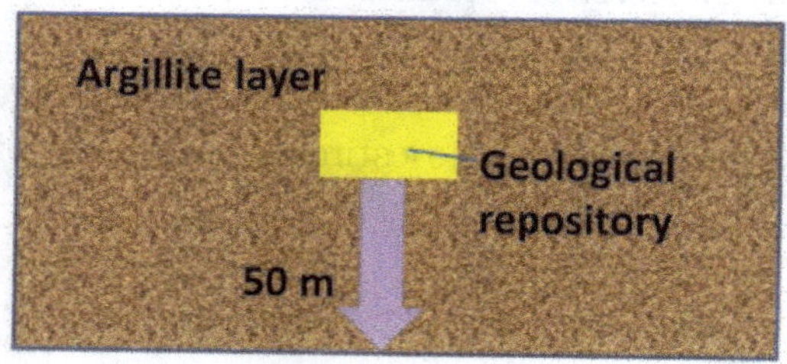

For question 6, the following parameters have been measured in the laboratory using representative samples of the argillite. Otherwise, they are considered reasonable estimates.

Sorption constant, K_d	10 mL/g
Porosity, e	0.3
Dry bulk density, ρ_b	2.66 g/cm^3
Effective porosity, η_e	12%
Saturated hydraulic conductivity, K_{sat}	1.3×10^{-5} cm/s
Dispersion coefficient, D	2.4×10^{-7} m^2/s

Directions and Hints

A. Go to GroundwaterSoftware.com at http://www.groundwatersoftware.com/calculator.htm
There are 17 on-line calculators. You will use three of them.

B. Go to Calculator 13, and calculate the retardation factor, R.

C. Go to Calculator 1 and calculate the average linear groundwater velocity.
Hint: You must enter the hydraulic conductivity in units of m/s.
Hint: The hydraulic gradient of vertical flow is 1.0 (unitless).

D. Go to Calculator 5 and calculate the concentration of ^{129}I after 100 years.
Hint: Distance is in units of meters, and time as seconds.
Hint: For groundwater velocity, insert the result from Calculator 1.

Bibliography

Busenberg, E., Plummer, L. N., and Parker, V. B. (1984). The Solubility of Strontianite ($SrCO_3$) in CO_2-H_2O Solutions between 2 and 91°C, the Association Constants of $SrHCO_{3+(aq)}$ and $SrCO_{3^0(aq)}$ between 5 and 80°C, and an Evaluation of the Thermodynamic Properties of $Sr^{2+}_{(aq)}$ and $SrCO_{3(cr)}$ at 25°C and 1 atm Pressure. *Geochimica et Cosochimica Acta*, 48, pp. 2021–2035.

Brookens, D. G. (1984). *Geochemical Aspects of Radioactive Waste Disposal*, Springer-Verlag Inc., New York, New York.

Cantrell, K. J. and Felmy, A. R. (2012). *Plutonium And Americium: A Site-Wide Review.* U.S. Department of Energy (Report RPT-DVZ-AFRI-001).

Cornell, R. M. (1993). Adsorption of Cesium on Minerals: A Review. *Journal of Radioanalytiucal and Nuclear Chemistry*, 171, pp. 483–500.

Fetter, C. W. (1993). *Contaminant Hydrogeology*, 2nd Ed.,, Prentice-Hall Inc., Upper Saddle River, New Jersey.

Follin, S. (2008). *Bedrock Hydrogeology Forsmark. Site Descriptive Modelling, SDM-Site Forsmark.* Swedish Nuclear Fuel and Waste Management Company (Report R-08-95).

Freeze, R. A. and Cherry, J. A. (1979). *Groundwater*, Prentice-Hall, New Jersey.

GSJ. (2005). *Atlas of Eh-pH Diagrams. Intercomparison of Thermodynamic Databases.* Geological Survey of Japan (Open File Report No. 419).

Lu, C. J., Liu, C. L., Chen, T., Wang, J., Wand, X. Y., Su, R., Sun, J. Y., Yang, R. X. and Khang, X. S. (2008). Determination of the Effective Diffusion Coefficient for $^{125}I^{-}$ in Beishan Granite. *Radiochimica Acta*, 96, pp. 111–117.

Lyman, W. J., Reidy, P. J., and Levy, B. (1992). *Mobility and Degradation of Organic Contaminants in Subsurface Environments.* C. K. Smoley, Inc., Chelsea, Michigan.

Missana, T., Garcia-Gutierrez, M., and Alonso, U. (2004). Kinetics and Irreversibility of Cesium and Uranium Sorption onto Bentonite colloids in a Deep Granitic Environment. *Applied Clay Science*, 26, pp. 137–150.

Ogata, A. (1970). *Theory of Dispersion in Granular Medium.* U. S. Geological Survey Professional Paper 411-l.

Roy, W. R., Krapac, I. G., Chou, S. F. J., and Griffin, R. A. (1992). *Batch-Type Procedures for Estimating Soil Adsorption of Chemicals.* U.S. Environmental Protection Agency (Technical Resource Document, U.S. EPA/530/-SW-87-006-F).

Siegel, M. D. and Bryan, C. R. (2003). *Environmental Geochemistry of Radioactive Contamination.* Sandia National Laboratories, Albuquerque, New Mexico (Report number SAND2003-2063).

Stephens, M. B., Follin, S., Peterson, J., Isaksson, H., Juhlin, C., and Simeonov, A. (2015). Review of the Deterministic Modelling of Deformation Zones and Fracture Domains at the Site Proposed for a Spent Nuclear Fuel Repository, Sweden, and Consequences of Structural Anisotropy. *Tectonophysics*, 653, pp. 68–94.

Wiles, D. R. (2002). *The Chemistry of Nuclear Fuel Waste Disposal*, Polytechnic International Press, Montreal, Canada.

Chapter 3

Uranium and Thorium Resources and Wastes

"My head bounded against a pointed rock, and I lost all knowledge of existence."

— Henry Lawson in *Journey to the Center of the Earth* by Jules Verne

3.1 Uranium

3.1.1 *Introduction*

Uranium is the basis for nuclear energy. The discovery of uranium is credited to a German chemist by the name of Martin Klaproth in 1789. Rather than naming the new element after himself, he chose to name it after the newly discovered planet Uranus. Uranium is an element with an atomic number of 92 that occurs in the Actinide Series of the Periodic Table. There are 28 known isotopes of uranium, yielding isotopes 215–242. Only three isotopes occur naturally: uranium-234 (0.0055% of all uranium), uranium-235 (0.72%), and uranium-238 (99.284%). Uranium-235 is fissile, meaning that it can sustain a nuclear reaction. Uranium-238 is fissionable by fast neutrons, and is fertile, meaning it can be transmuted into fissile plutonium-239 in a nuclear reactor.

Because uranium is mostly composed of ^{238}U, a hand-held sample is only weakly radioactive because ^{238}U decays to ^{235}Th by alpha decay with a long half-life that is comparable to the age of the earth (circa 4.54 Ga). Isotopes 215–228 are less stable, and have short

Table 3.1. Uranium isotope numbers 229–242 (derived from Audi *et al.*, 2003; Devaraja *et al.*, 2016; Meija *et al.*, 2016, and references therein).

Isotope number	Half-life	Decay mode	Immediate daughter-decay product
229	58 min	β^+ and α	^{229}Pa and ^{225}Th
230	20.8 days	α	^{226}Th
231	4.1 days	Electron capture	^{231}Pa
232	68.9 years	α	^{228}Th
233	1.592×10^5 years	α	^{229}Th
234	2.455×10^5 years	α	^{230}Th
235	7.04×10^8 years	α	^{231}Th
236	2.342×10^7 years	α	^{232}Th
237	6.75 days	β^-	^{237}Np
238	4.469×10^9 years	α	^{234}Th
239	23.5 min	β^-	^{239}Np
240	14.1 hs	β^-	^{240}Np
241	about 5 min	β^-	^{241}Np
242	about 16.8 min	β^-	^{242}Np

half-lives ($\leq$9.1 min), and also decay to thorium as

$$^{n}U \rightarrow \alpha + {}^{n-4}Th \tag{3.1}$$

The properties of isotopes 229–241 are summarized in Table 3.1.

3.1.2 *Global Abundance and Mining*

The World Nuclear Association (2019) estimated that the world's known resources of uranium are 6,142,600 tonnes U. About 30% of these resources are in Australia, followed by Kazakhstan (14%) and the Russian Federation and Canada (8%). World mining production in 2018 was led by Kazakhstan, Canada, and Australia accounting for 65.8% of all the uranium produced. Mining methods include underground mines, open pits, *in situ* leaching, and heap leaching. The Cigar Lake Mine is an underground mine in Saskatchewan in Canada. It produced about 13% of the world's demand for uranium in

Figure 3.1. Open-pit uranium mine at the McClean Lake facility in Canada (Canadian Nuclear Safety Commission). Available at: http://www.nuclearsafety.gc.ca [Accessed 17 January 2020]. Used with permission of the Canadian Nuclear Safety Commission.

2018 (World Nuclear Association, 2019). McArthur River Mine had produced uranium from deposits that are at a depth of between 500 and 600 m (Jamieson, 2002). The average concentration of uranium in the ore is 15% as U_3O_8 with some drill cores encountering ore containing as much as 43% of U_3O_8. The Ranger Uranium Mine in northern Australia is the world's second largest mine, and it consisted of three ore bodies when mining began in 1980. It is an open-pit mine. Overburden is first removed at the surface until the uranium ore is accessible. The process creates a large crater, depending on the depth and distribution of the ore bodies (Fig. 3.1). In 2018, open pits and underground mines accounted for 39% of the world demand for uranium (WNA, 2019).

The third mining technique is called *in situ* leaching, *in situ* recovery, or solution mining. It is usually applied to low-grade ores that are too deep to justify near-surface mining (Bhargava *et al.*, 2015). Injection wells are placed into uranium-containing material, and sulfuric acid is pumped down to dissolve the uranium (Fig. 3.2). The dissolved uranium is then recovered using extraction wells and

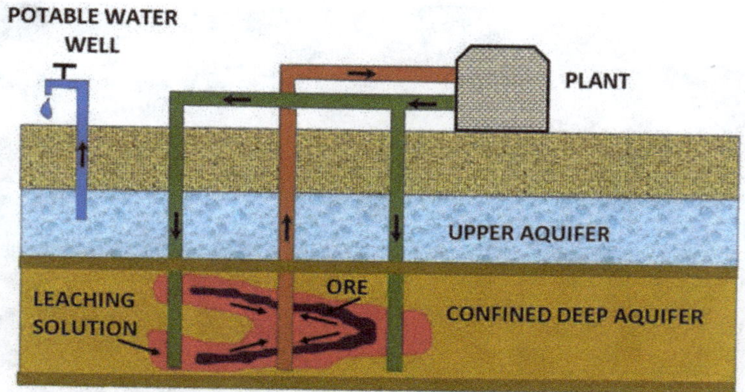

Figure 3.2. ***In situ*** **leaching (Wise Uranium Project) Available at: http://www.wise-uranium.org/uisl.html [Accessed 17 January 2020]. Used with permission of the Wise Uranium Project.**

brought to the surface for refinement. Ideally, the uranium-rich layer is bounded by relatively water-impermeable layers that prevent the contamination of groundwater by the acidic, uranium-rich fluid. *In situ* leaching is used in Kazakhstan. In 2018, it accounted for 55% of the world demand for uranium (World Nuclear Association, 2019).

A fourth technique is heap leaching. This approach is used for extracting uranium from low-grade ore. The ore is first excavated, crushed, and then piled on a leaching pad with either a plastic or compacted soil liner (Fig. 3.3). Sulfuric acid or an alkaline extractant is added to the top of the pile and allowed to percolate down until it reaches the liner. The pile may be tilted to facilitate gravity drainage into a collection basin. The extracted uranium will be refined to ultimately produce yellow cake. Fresh ore may then be placed on the leach pad. Additional information can be found in McBride *et al.* (2012).

The mineralogical composition of the ore can impact the efficacy of leaching. Carbonate minerals, for example, may neutralize the sulfuric acid rather than the acid dissolving the uranium-containing minerals. Heap leaching studies have been conducted in Argentina, Canada, France, Spain, and Australia (Bhargava *et al.*, 2015).

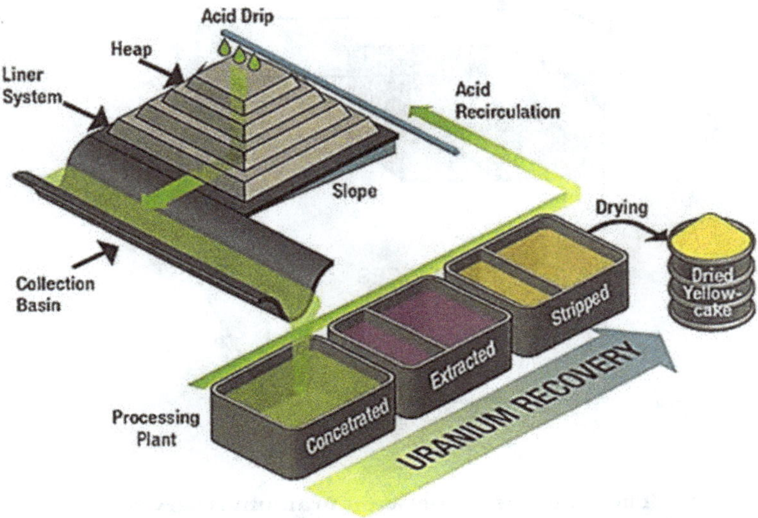

Figure 3.3. Heap leaching of uranium ore (US NRC, 2015). Available at: https://www.nrc.gov [Accessed 17 January 2020].

3.1.3 *Uranium Quality*

The term "ore" is an economic term. It is applied to a geological body such as one containing uranium in which the concentration of the metal is large enough to economically justify mining. Uranium reserves are those of known volume and concentration of uranium and are economical to mine. Uranium resources that are suspected to be present based on geological inference (Fig. 3.4). Resources include undiscovered sources in addition to ones that are too deep to mine, or those occurring in insufficient amounts to mine because of prevailing market conditions.

Secondary sources is a broad term which includes uranium from reprocessing used nuclear fuel, and feed stocks created by the decommissioning of Cold War nuclear weapons. Unconventional resources include uranium-rich coal ash. Some coal seams in the Yunnan Province of China, for example, contain anomalously large

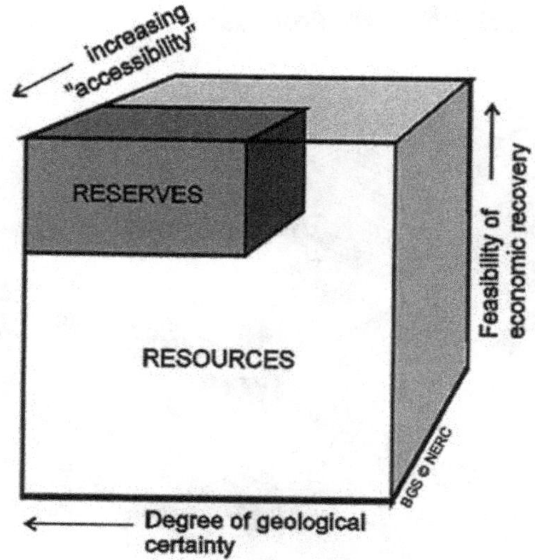

Figure 3.4. The relationship between uranium reserves and resources (Minerals UK, 2017). Available at: http://www.bgs.ac.uk/mineralsUK [Accessed 17 January 2020]. Used with permission of the British Geological Survey.

amounts of uranium. Experiments are on-going to extract the uranium from the ash. Some black shales in the US contain relatively large amounts of uranium, but few were economical to mine (see Swanson, 1961).

Sea water is another interesting source of uranium. The concentration of uranium in sea water is only about 3.3 μg/L but yielding an estimated 5×10^9 tonnes of uranium in the oceans. Progress has been made in extraction uranium from sea water, and the process might be practical and economical in the future. Current field-scale research involves submerging stationary, uranium absorbents, followed by extracting the uranium. A recent review was provided by Dungan *et al.* (2017).

3.1.4 *Geological Modes of Occurrence of Uranium*

Uranium is fairly ubiquitous at background levels in geomedia. It ranges from about 0.3 to 11 mg/kg in soils (Shacklette and Boerngen,

1984). Morrow (2001), for example, reported that the concentration of uranium in the groundwater in Illinois varied from <0.1 to 17 μg/L. There are also several geochemical processes such as the dissolution of uranium followed by chemical precipitation that have the capacity to concentrate background levels in other areas away from the source. Uranium is mobile under oxidized conditions, and precipitates under reduced conditions.

The main sources of uranium are hydrothermal vein deposits (Bhargava *et al.*, 2015). The genesis of this type of deposit begins with the intrusion of a hydrothermal solution which is subsurface, high-temperature groundwater. Heated by subsurface magma bodies, the heated groundwater rises to the surface where the water may cool, and minerals may precipitate along the walls of fractures and faults (Fig. 3.5). If these dissolved mineral phases contain uranium, then

Figure 3.5. Hydrothermal vein deposits in Washington, US (Geology Café.com, 2015. Available at: http://geologycafe.com [Accessed 17 January 2020]. Used with permission from Geology Café.

hydrothermal solutions can transport uranium toward a near-surface environment that is accessible to mining.

Unconformity-related deposits are the major sources of uranium in Canada and Australia. An unconformity is a sequence of rock layers in which a layer appears to be missing between the older rocks below, and younger rocks directly above. One mechanism to account for the absent layer is that the surface of older material was eroded before the deposition of younger material. In the context of this chapter, the unconformity may be separated by residual material in which uranium has been concentrated. Sometimes, the residual material is a breccia which is a rock composed of broken fragments of rocks and minerals. If the source of the breccia contained uranium, the accumulation of breccia may become an ore of uranium.

Sandstone–uranium deposits are another important source of uranium. Uranium-rich groundwater can flow through porous sandstone, and depending on the geochemical conditions present, the uranium can precipitate and be concentrated as solid, uranium-containing phases. Other uranium deposits include uranium-rich metamorphic rocks, quartz-pebble conglomerates (a weathering residue that can be transported and concentrated), and metasomatic deposits (rocks chemically altered by hydrothermal solutions). Additional information about other types of deposits were given by Bhargava *et al.* (2015).

3.1.5 *Uranium Minerals*

Uranium occurs in more than 200 minerals (Bhargava *et al.*, 2015), but the major ore of uranium is called uraninite (Fig. 3.6). Its chemical composition is often given as UO_2. It was formerly called pitchblende which is written as $U_2O_5 \cdot UO_3$ or U_3O_8 to account for the two different oxidation states of uranium (U(IV) and U(VI)). Uraninite and pitchblende are often used as synonyms. A secondary mineral is one that forms after weathering reactions or by chemical reactions with hydrothermal solutions. Various secondary uranium

(a) (b)

Figure 3.6. Uraninite (a) and uraninite plus yellow alteration products (b). Available at: https://www.minerals.net/Image/10/143/Uraninite.aspx [Accessed 17 January 2020]. Used with permission of Minerals.net.

(a) (b)

Figure 3.7. Carnotite (a) and torbernite (b). Available at: https://en.wiktionary.org [Accessed 17 January 2020]. Used with permission of Wikimedia Foundation.

minerals (Fig. 3.7) occur by chemically combining the uranyl ion ($(UO_2)^{2+}$) with various cations and anions (Table 3.2).

3.1.6 *The Longevity of Uranium as a Fuel*

There is considerable uncertainty about how long the supply of uranium will be available to sustain nuclear energy. For example,

Table 3.2. Selected secondary uranium minerals.

Mineral	Chemical composition
Autunite	$Ca(UO_2)_2(PO_4)_2 \cdot 10 - 12H_2O$
Brannerite[a]	UTi_2O_6
Carnotite	$K_2(UO_2)_2(VO_4)_2 \cdot 1 - 3H_2O$
Torbernite	$Cu(UO_2)_2(PO_4)_2 \cdot 12H_2O$
Tyuyamunite	$Ca(UO_2)_2(VO_4)_2 \cdot 5 - 8H_2O$
Uranophane	$Ca(UO_2)_2(HSiO_4)_2 \cdot 5H_2O$

[a]Relatively insoluble in sulfuric acid, and it is not a major source of uranium.

Day (1975) predicted that the supply of uranium ore would be exhausted by 2015. In 2001, the European Commission concluded that, at the current rate of consumption, the known resources of uranium would be exhausted by 2043. If secondary sources were included, the Commission extended the year of depletion to about 2073 (*The Sunday Times* [London] 2001). The former Australian Uranium Association predicted that the world's "present measured resources" would be depleted by about 2079 (World Nuclear Association, 2019).

A report issued by NEA–OECPD (2001) proposed that all known sources of uranium could be depleted by 2087. If all undiscovered and unconventional sources were included, it was predicted that uranium could be available until 2273. If breeder reactors become practical, the NEA–OECPD predicted that uranium could sustain nuclear energy for 8,500 years. An even more optimistic prediction was given by NEA–IAEA (2016). They proposed that, if all known resources, and secondary reserves are used, uranium should be available for about 47,000 years. Lastly, Cohen (1983) argued that if fast reactors and extracting uranium from sea water both become practical, the supply of uranium could last for as long as 5 billion years.[1]

[1]None of the longevity estimates seemed to explicitly consider the potential impact of transmuting thorium-235 into uranium-233.

Radioactive Dinnerware

The Fiesta dinnerware line was introduced in 1936. The orange colors associated with the plates, bowls, and cups were created by adding uranium oxide to the glaze. The orange dinnerware can easily set off a Geiger counter which is a popular classroom demonstration. The addition of uranium oxide was discontinued in 1944. In 1959, a new line of orange Fiesta ware was made by adding depleted uranium to the glaze. Ten years later, the production of orange Fiestaware using depleted uranium ceased.

3.1.7 *Uranium-Related Wastes*

During the mining of uranium ore, the overburden (soil and rock materials) must be removed to allow access to the ore. In the case of open pit mining, the ratio of the volume of waste rock to the volume of the ore may be 40:1 or more (Frost, 1998). Underground mining results in tunnels to reach the ore, and the ratio of waste rock to ore may be less than one. The decision about pit mining versus underground mining depends on the depth of the ore, rock stability (to support tunnels), economic considerations, and safety factors.

The chemical and mineralogical composition of the waste rock is site-specific and depends on the geological history of the area. In general, however, the waste rock may contain uranium in concentrations that are less than those needed to classify the overburden as an ore. When uranium is present, its daughter decay products will also be present, including radon. Another environmental concern is the presence of sulfide minerals such as pyrite (FeS_2), marcasite (FeS_2), and sphalerite ((Zn, Fe)S). After excavation, the sulfide minerals in the waste rock may oxidize when it is piled on the surface and form sulfuric acid which may in turn leach potential contaminants such as arsenic, lead and radionuclides associated with uranium from the waste rock. The generation of sulfuric acid is not unique to uranium-containing rocks but can occur during the mining of other minerals such as lead, copper, and coal.

Frost (1998) generalized that, depending on the acid-generating potential of the waste rock, less problematic material may be left on the surface, graded, covered with soils, and re-vegetated. Waste rock that has a significant acid-generating potential is used for mine backfill or placed into mined-out pits. Diehl (2011) described European experiences in which the waste rock has been crushed into gravel, and used in cement, road construction, and railroad track foundations.

Uranium mill tailings are another type of waste. They are the byproduct of extracting uranium from the ore. The ore is crushed and leached with sulfuric acid or an alkaline extractant. The resulting waste tailings tend to occur as a sludge that is stored in surface impoundments or as piles (Fig. 3.8). Additional information was given by Diehl (2011).

Depleted uranium is a byproduct of uranium enrichment. It is depleted in the sense that it contains about 0.3% ^{235}U. There is a

Figure 3.8. Uranium mill tailings in Saskatchewan, Canada (Canadian Nuclear Safety Commission, 2020. Available at: https://nuclearsafety.gc.ca [Accessed 19 January 2020]. Used with Permission of the Canadian Nuclear Safety Commission.

debate whether depleted uranium is a waste or a resource. It has been used in counterweights because of its density, but more often in the manufacturing of armor-piercing ammunition. Current US NRC regulations for waste classification do not address depleted uranium, although it resembles a Class A low-level waste (see Chapter 4). The status and future of depleted uranium for waste management are currently under study (https://www.nrc.gov/waste/llw-disposal/llw-pa/uw-streams/bg-info-du.html).

3.2 Thorium

3.2.1 *Introduction*

Thorium is a potential fuel for nuclear energy. The Swedish chemist Jons Jacob Berzelium discovered thorium in 1828. He named the new discovery after Thor, the hammer-wielding Norse god of thunder and lightning. Thorium is an element with an atomic number of 90 that occurs in Period 7 of the Periodic Table. There are 32 known isotopes of thorium, yielding a range in isotopes from atomic mass 209 to 238, and all of the isotopes are radioactive. All of the isotopes less than 225 atomic mass units have half-lives that are less than 1.5 s. In terms of abundance, 99.98% of all thorium occurs as ^{232}Th. The half-life of ^{232}Th is 1.405×10^{10} years which is longer than the age of the Earth. It first decays to mostly ^{228}Ra by alpha decay. About 0.02% of all thorium occurs as ^{230}Th which has a half-life of 75,400 years. It decays to ^{226}Ra by alpha decay. The remaining five naturally occurring isotopes exist in trace amounts and are ^{227}Th, ^{228}Th, ^{229}Th, ^{231}Th, and ^{234}Th. All of the isotopes greater than 234, have a half-life that is less than 10 min. The properties of thorium isotopes 225–234 are summarized in Table 3.3.

3.2.2 *Global Abundance and Geological Modes of Occurrence*

The World Nuclear Association (2017) estimated that the world's known resources and inferred reserves of thorium to be about 6,355,000 tonnes which is greater than those of uranium. The four

Table 3.3. Thorium isotope numbers 225–234 (derived from Audi *et al.*, 2003; Meija *et al.*, 2016, and references therein).

Isotope number	Half-life	Decay mode	Immediate daughter-decay product
225	8.72 min	α, EC	^{221}Ra
226	30.57 min	α	^{222}Ra
227	18.68 days	α	^{223}Ra
228	1.911 years	α	^{224}Ra
229	7,917 years	α	^{225}Ra
230	75,400 years	α	^{226}Ra
231	25.53 h	β^+	^{231}Pa
232	1.405×10^{10} years	α	^{228}Th
233	21.83 min	β^-	^{233}Pa
234	24.30 days	β^-	^{234m}Pa

Figure 3.9. Monazite.

major countries with thorium are India (with 13% of the total), Brazil (10%), Australia and the USA (both 9%), and Egypt (6%).

The primary source of thorium is a phosphate mineral called monazite ((Ce, La, Nd, Th)(PO_4, SiO_2)) (Fig. 3.9). There are other thorium-containing minerals. They are, however, rare and include euxenite ((Y, Co, Ce, U, Th) (Nb, Ta, Ti)$_2O_6$), and thorianite (ThO_2) (Fig. 3.10).

The average content of thoria in deposits in Sri Lanka was reported as 14.3%, 5.9% in South Africa, and 3.1% in the USA (IAES, 2005). India and China are currently the major producers of

(a) (b)

Figure 3.10. Euxenite (a). Available at: https://www.mindat.org/min-1425.html and thorianite (b). Available at: https://www.weinrichmineralsinc.com/products/thorianite-2501316.php [Both accessed 19 January 2020]. Used with permission from Weinrich Minerals.

monazite. Malaysia, Vietnam, and Brazil are also mining monazite. Neither the US nor Australia are currently mining the phosphate mineral. The global resources of monazite were estimated to be about 12 million tonnes.

Monazite forms during the crystallization of igneous rocks such as granitic pegmatites, and during the metamorphism of some sedimentary rocks. As these source rocks become exposed at the Earth's surface, they weather by physical, chemical, and biological processes. Monazite is relatively resistant to weathering and accumulates as a residual mineral while more the labile minerals are dissolved or altered. Depending on the prevalent geological conditions, the monazite may be transported by gravity and/or water and then be deposited at a location removed from the sources. The resulting monazite accumulation may be classified as a placer deposit which is a natural accumulation of valuable minerals in alluvium (river deposits), colluvium (material at the bottom of an incline), beach sands, and wind deposits.

3.2.3 *Thorium as a Potential Fuel*

The potential to develop and use the Thorium Fuel Cycle has been debated for decades. There was an interest in thorium from circa

Table 3.4. The potential advantages and disadvantages of the Thorium Fuel Cycle.

Advantages	Disadvantages
1. Thorium is more abundant than uranium in nature.	1. Thorium is not fissile: it cannot sustain a chain reaction.
2. Thorium-containing minerals (primarily monzanite) are more easily mined in many countries than uranium.	2. Thorium must be transmuted into ^{233}U to be used as a fissile fuel.
3. Thorium has the potential to supplement the supply of uranium.	3. Uranium-233 must be extracted from irradiated thorium-232.
4. In the ^{232}Th–^{233}U fuel cycle, plutonium is not produced, and fewer actinides are created.	4. Thorium-232 will need to be mixed with a "driver" such as ^{233}U, ^{235}U or ^{239}Pu.
5. The presence of ^{232}U could create desirable non-proliferation properties.	5. Because of the presence of ^{232}U, thorium-based fuels may need to be processed remotely in a shielded environment.
	6. Thorium-based fuels are currently more expensive than uranium-based fuels.
	7. Other than in India, there is little global interest to develop thorium-based fuels, and experiences have been negative and limited.

1950 to 1970 on a global scale, but this interest faded by the 1980's in most countries with the exception of India (Lung and Gremm, 1998). However, there are still thorium advocates who discuss its use as an alternative to uranium with optimism and enthusiasm (see, for example, Hargraves, 2012; Martin, 2012; Sehgal, 2013). Other authors have been less optimistic (NNL, 2010).

Table 3.4 presents a summary of the advantages and disadvantages of the Thorium Fuel Cycle, based on information provided by IAEA (2005), Lung and Gremm (1998), and WNA (2017). It is beyond the scope of the chapter to persuade the reader to join or reject the thorium bandwagon. A careful study of Table 3.4 will, however, serve as a useful basis for discussion for the rest of this chapter. Starting with items 6 and 7 in Table 3.4, thorium must be transmuted into a fissile radionuclide, namely thorium-233 by absorbing a neutron:

$$^{232}\mathrm{Th} + \eta \rightarrow {}^{233}\mathrm{Th} \rightarrow {}^{233}\mathrm{Pa} + \mathrm{e}^- + \underline{\nu} \tag{3.2}$$

$$\text{followed by } {}^{233}\mathrm{Pa} \rightarrow \mathrm{e}^- + \underline{\nu} + {}^{233}\mathrm{U} \tag{3.3}$$

where η is a neutron, Pa is the protactinium, e^- is the beta particle, and $\underline{\nu}$ is the antineutrino.

3.2.4 *Undesirable Side Reactions*

Thorium-233 has a half-life of only 21.8 min, and protactinium-233 has a half-life of 27.0 days to yield uranium-233 (Peterson *et al.*, 2007). The transmutation of ^{232}Th also yields ^{232}U. Because ^{233}U and ^{232}U differ by one neutron, their chemical reactivity is identical. They cannot be separated chemically. Both are fissile, but only ^{233}U has been used as a fuel in a reactor in India (IAEA, 2005). Uranium-233 has a relatively long half-life of about 159,000 years, and decays into ^{229}Th by alpha decay (IAEA, 2005). In contrast, uranium-232 has a much shorter half-life of 73.6 years, followed by a day sequence that includes some "strong gamma emitters" (Lung and Gremm, 1998). Most of the decay products have short half-lives (Fig. 3.11). Toward the end of the decay chain are the sources of significant gamma radiation. Lead-212 yields 0.15 MeV, ^{212}Bi releases 0.19 MeV, and ^{208}Tl emits 3.4 MeV of gamma radiation (Peterson *et al.*, 2007).

The presence of these significant gamma radiation sources has been interpreted as both a blessing and a curse. A potential advantage could be that such radiotoxic decay products could increase the non-proliferation properties of thorium-based fuel and wastes (item 5 in Table 3.4). The presence of the same radiotoxic products could also be a disadvantage because utilizing thorium-based fuel may require remote handling and additional shielding (item 10) which in turn

$^{232}U \rightarrow {}^{228}Th \rightarrow {}^{224}Ra \rightarrow {}^{220}Rn \rightarrow {}^{216}Po \rightarrow {}^{212}Pb \rightarrow {}^{212}Bi$ . . .
α α α α α α
69 years 1.9 year 3.7 days 56 sec 0.15 sec 11 hours

(64% time)
$^{212}Bi \rightarrow {}^{212}Po \rightarrow {}^{208}Pb$
β^- α stable
61 minutes 0.3 μsec

(36% time)
$^{212}Bi \rightarrow {}^{208}Tl \rightarrow {}^{208}Pb$
α β^- stable
61 minutes 3 minutes

Figure 3.11. The uranium-232 decay chain.

would increase the costs of fuel fabrication when compared with those of uranium-based fuels (item 11).

3.2.5 *Thorium Drivers*

Most previous reactor experiments with thorium have focused on adding fissile isotopes to thorium as "drivers" to initiate and sustain a nuclear reaction. These drivers have included ^{233}U (extracted from irradiated thorium-232), ^{235}U, and ^{239}Pu. IAEA (2005) summarized the reactor type, the fuel configuration or shape, and the fuel element that contained the fuel shape.

The fuel shapes have included microspheres (200–800 μm in diameter), sintered pellets, and molten salt. It is beyond the scope of this chapter to discuss all these thorium-based fuel designs. The basic design most often used to combine thorium with fissile drivers used spherical-fuel elements called pebbles that were composed of a pyrolytic graphite (Fig. 3.12). Within each pebble were micro-fuel particles composed of layers of pyrolytic carbon and silicon carbide. Within the center of each particle was a mixture of thorium and uranium (written as (Th, U) O_2), or plutonium ((Th, Pu) O_2)). The center was called the "kernel" and was embedded in a graphite mixture.

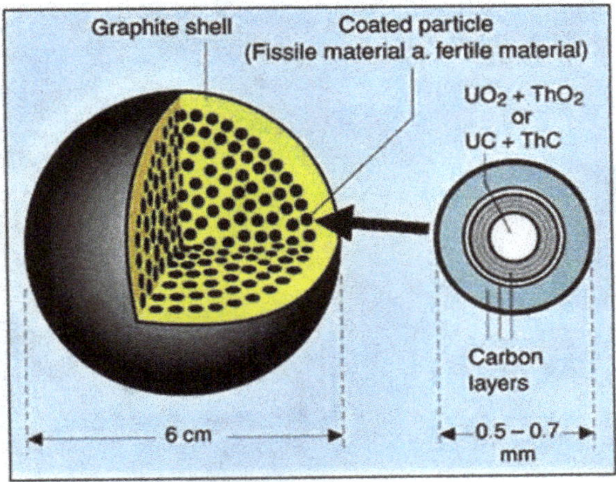

Figure 3.12. Coated particle (right) and a pebble (left) (European Nuclear Society. Available at: https://www.euronuclear.org/info/encyclopedia/p/pebble.htm [Accessed 19 January 2020]. Used with permission of the European Nuclear Society.

THOREX

If the recovery of ^{233}U from irradiated thorium or thorium-based fuels is required, it can be extracted chemically. Uranium-233 has been recovered from spent thoria assemblies in India (IAEA, 2005). The most viable procedure available is called THOREX (THORium-uranium Extraction). Thoria is relatively insoluble in water and resistant to dissolution in strong acids. The crushed or ground spent thoria is boiled in a solution called the THOREX reagent. It is composed of 13 M HNO_3 + 0.05 M HF + 0.1 M $Al(NO_3)_3$. Similar to the PUREX procedure, tributyl phosphate and dodecane are used to extract the uranium and thorium while the fission products remain in the aqueous phase. The co-extracted thorium can be stripped from the organic phase by adding nitric acid to the organic phase. The ^{233}U can then be collected using cation exchange columns.

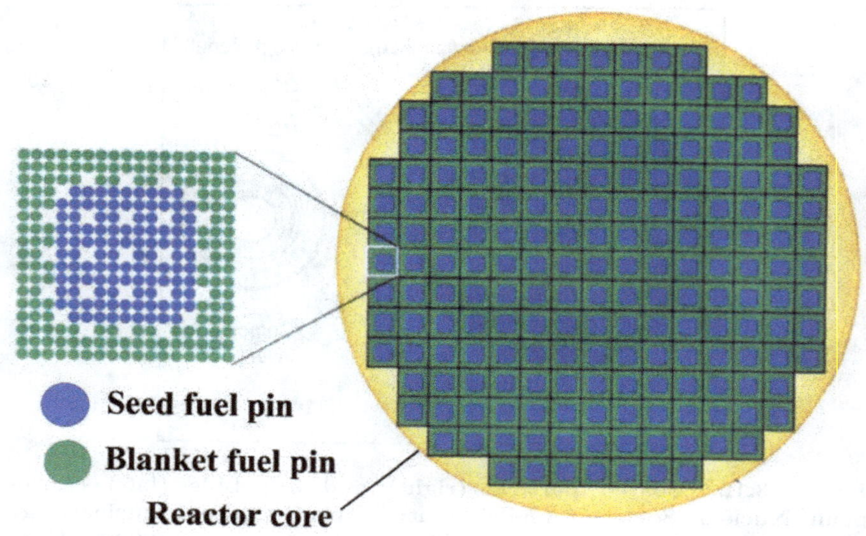

Figure 3.13. The seed (fissile fuel) and blanket (thorium or thoria) pins (Sehgal, 2013). Used with permission of CERN.

Another fuel design that uses thorium with a fissile material is called the seed-blanket fuel design. It is also known as the Radkowsky Thorium fuel design, named after Alvin Radkowsky, a nuclear engineer who pioneered the concept. Basically, the seed is composed of ^{235}U, ^{233}U, Pu, or mixed-oxide fuel (MOX) fuel. The blanket is composed of Th or thoria. Within each fuel assembly, seed rods are surrounded by the blanket rods (Fig. 3.13).

3.2.6 *Thorium-based Spent Fuel*

There is very little information available on the radionuclide concentrations in thorium-based spent fuel. IAEA (2005) listed 18 nuclear reactors that used thorium. Many were short-lived experiments that were designed and operated to test different reactor designs, fuel combinations, and fuel designs. These reactors were operated in Canada, Germany, India, the Netherlands, the UK, and the US. The majority of these experiments were terminated by 1989. The reactors were shut-down for a variety of reasons such as fuel element problems, water and gas leaks, groundwater contamination, corrosion issues, a loss of funding, and a loss of support and interest in thorium-based nuclear energy.

Common to all but one of these case studies (the Peach Bottom Power Plant in the US) was a lack of any investigations on the post-exposure isotopic composition of the spent fuel. Either such spent fuel characterization was not a research priority, or the data were not made available to the public. The Peach Bottom Power Plant is the only, documented source of data on the chemical composition of thorium-based spent fuel.

The Peach Bottom Unit 1 reactor was an experimental reactor that was operated in Pennsylvania (US) from 1966 to 1972. Unit 1 used coated micro-particles that were between 90 and 130 μm in diameter (Morissette *et al.*, 1986). The kernels contained ^{235}U that was 93% enriched. Details about the initial thorium or uranium concentrations in the kernels were not reported. The ratio of uranium to thorium was reported as 0.2 and 0.6, depending on the fuel element. Two core samples were collected after the reactor was shut down. Details about how the cores were collected, sampled, and analyzed were not provided. The partial radioisotopic composition of Core 2 was estimated from the data in Morissette *et al.* (1986) and is given in Table 3.5. Thorium-232 was the dominant residual fuel component. Uranium-233 formed as a transmutation product of ^{232}Th. Because ^{235}U was initially present in the kernels, plutonium isotopes would be expected in the spent fuel. The compositional data in Table 3.5 may not be representative of that produced in future commercial applications. The data do, however, provide a limited insight on a thorium-based waste product.

It should be noted that a mini-reactor in India (KAMINI) uses only ^{233}U as a fuel and is still in operation. However, there is no information available on the chemical composition of the spent fuel. Because of a lack of actual thorium-based fuel samples, recent reports have been based on computer simulations. For example, Galaholm (2017) applied the MCNPX code (Monte Carlo N-Particle code that models the interactions of radiation with matter) that was developed by Los Alamos (US). It was applied to six types of end-of-life fuels,

Table 3.5. Partial isotopic composition of spent thorium-based fuel from the Peach Bottom Unit 1 reactor.

Radionuclide	Concentration	Probable origin
^{232}Th	90%	Residual fuel component
^{238}U	0.7%	Residual fuel component
^{236}U	1.6%	Residual fuel component
^{235}U	5.2%	Residual fuel component
^{231}Pa	4.5 mg/kg	Decay product of ^{231}Th
^{233}Pa	0.02%	Decay product of ^{233}Th
^{232}U	5.8 mg/kg	Product of ^{232}Th transmutation
^{233}U	2.0%	Product of ^{232}Th transmutation
^{237}Np	0.12%	Decay product of ^{237}U
^{239}Pu	153 mg/kg	Formed by neutron capture by ^{238}U
^{240}Pu	53 mg/kg	Formed by neutron capture by ^{239}Pu
^{241}Pu	87 mg/kg	Formed by neutron capture by ^{240}Pu
^{242}Pu	41 mg/kg	Formed by neutron capture by ^{241}Pu

Note: The concentrations were estimated from the Core 2 data in Morissette *et al.* (1986).

based on the nuclear reactor design VVER-1200 (a water–water energetic reactor originally designed in the former Soviet Union). The fuel types were UO_2 (alone as a basis for comparison), ^{232}Th + ^{235}U, ^{232}Th + ^{233}U and Th + U + Pu. The model simulations indicated that plutonium, americium, neptunium, and cerium would not be present in the spent fuel when ^{233}U was combined with ^{232}Th. The presence of Pu was predicted in all the other combinations. It appears that when ^{235}U is used as a trigger fuel, plutonium will be present in the spent fuel which argues against the non-proliferation advantages of using the Thorium Fuel Cycle.

3.3 Review Questions

1. How many isotopes of uranium are known?
2. Which country has the largest uranium resources?
3. Compare *in situ* leaching with heap leaching.
4. What is the difference between a resource and a reserve?
5. How long will uranium last as a fuel for nuclear energy?

6. What is the difference between mining rock waste and uranium mill tailings?
7. What is the major isotope of thorium?
8. What is the major thorium-containing mineral?
9. Name three potential advantages of the Thorium Fuel Cycle when compared with uranium-based fuels.
10. Why is the presence of ^{232}U in thorium spent fuel a potential barrier to the Thorium Fuel Cycle?
11. Define a pebble, kernel and blanket in the Thorium Fuel Cycle.
12. What is a "thorium driver" and how can it lead to the formation of plutonium?

Bibliography

Audi, G., Bersillon, O., Blachot, J., and Wapstra, A. H. (2003). The NUBASE Evaluation of Nuclear and Decay Properties. *Nuclear Physics A*, 729, pp. 3–128.

Bhargava, S. K., Ram, R., Pownceby, M., Grocott, S., and Ring, B. (2015). A Rreview of Acid Leaching of Uraninite. *Hydrometallurgy*, 151, pp. 10–24.

Cohen, B. L. (1983). Breeder Reactors: A Renewable Energy Source. *American Journal of Physics*, 51, pp. 75–76.

Day, M. C. (1975). Nuclear Energy: A Second Round of Questions. *Bulletin of the Atomic Scientists*, 31, pp. 52–59.

Devaraja, H. M., Heinz, S., Beliuskina, O., Comas, V., Hofmann, S., Hornung, C., Munzenberg, G., Nishio, K., Ackermann, D., Gambhir, Y. K., Gupta, M., Henderson, R. A., Hebberger, F. P., Khuyagbaatar, J., Kindler, B., Lommel, B., Moody, K. L., Manurer, J., Mann, R., Popeko, A. G., Shaughnessy, D. A., Stoyer, M. A., and Yeremin, A. V. (2015). Observation of New Neutron-Deficient Isotopes with Z $\geq$ 92 in Multinucleon Transfer Reactions. *Physics Letters B*, 748, pp. 199–203.

Diehl, P. (2011). Uranium mining and milling wastes: An introduction. Wise Uranium Project, Mining and Milling, Impacts. Available at: www.wise-uranium.org [Accessed 19 January 2020].

Dungan, K., Butler, G., Livens, F. R., and Warren, L. M. (2017). Uranium from Seawater–Infinite Resources and Improbable Aspiration? *Progress in Nuclear Energy*, 99, pp. 81–85.

Frost, S .E. (1998). *Waste Management in the Uranium Mining Industry*. The Uranium Institute, Twenty-Third Annual International Symposium 1998. Available at: www.world-nuclear.org/sym/1998/chambe.htm [Accessed 19 January 2020].

Galahorn, A. A. (2017). Minimization of the Fission Product Waste by Using Thorium Based Fuel Instead of Uranium Oxide. *Nuclear Engineering and Design*, 314, pp. 165–172.

Hargraves, R. (2012). *Thorium.* Create Space Independent Publishing Platform, Scotts Valley, Californium.

IAEA. (2005). *Thorium Fuel Cycle — Potential Benefits and Challenges.* International Atomic Energy Agency (Report number IAEA-TECDOC-1450).

Jamieson, B. W. (2002). Mining the High-Grade Mcarthur River Uranium Deposit. In *The Uranium Production and the Environment, International Symposium*, International Atomic Energy Agency, October 2–6, 2000, Vienna, Austria, pp. 272–286.

Lung, M. and Gremm, O. (1998). Perspectives of the Thorium Fuel Cycle. *Nuclear Engineering and Design*, 180, pp. 133–146.

Martin, R. (2012). *Super Fuel.* Palgrave MacMillan, New York.

McBride, D., Cross, M., and Gebhardt, J. E. (2012). Heap Leach Model Employing CFD Technology: A Process Heap Model. *Minerals Engineering*, 33, pp. 72–79.

Meija, J., Coplen, T. B., Berglund, M., Brand, W. A., Bievre, P.D., Groning, M., Holden, N. E., Irrgeher, J., Loss, R. D., Walczyk, T., and Prohaska, T. (2016). Atomic Weights of the Elements 2013. *Pure and Applied Chemistry*, 88, pp. 265–291.

Morissette, R. P., Tomsio, N., and Razvi, J. (1986). *Characterization of Peach Bottom Unit 1 fuel.* Oak Ridge National Laboratory (Report number ORNL/Sub/86-22047/2 GA-C18525).

Morrow, W. S. (2001). *Uranium and Radon in Ground Water in the Lower Illinois River Basin.* Water-Resources Investigations (Report 2001-4056).

NNL. (2010). *The Thorium fuel cycle.* The National Nuclear Laboratory (United Kingdom). Position Paper. Warrington, UK.

NEA-IAEA. (2016). *Uranium 2016: Resources, Production and Demand.* Nuclear Energy Agency–International Atomic Energy Agency. (NEA report 7301).

NEA-OECPD. (2001). *Uranium 2001: Resources, Production and Demand.* Nuclear Energy Agency–Organization for Economic Co-operation and Development.

Peterson, J., MacDonell, M., Haroun, L., Monette, F., Hilderbrand, R. D., and Taboas, A. (2007). *Radiological and Chemical Fact Sheets to Support Health Risk Analyses for Contaminated Areas.* Argonne National Laboratory.

Sehgal, B. R. (2013). Feasibility and Desirability of Employing the Thorium Fuel Cycle for Nuclear Power Generation. In *Thorium Fuel: THEC 13*, Geneva, Switzerland, October 28–31, 2013.

Shacklette, H. T. and Boerngen, J. G. (1984). *Element Concentrations in Soils and Other Surficial Materials of the Conterminous United States.* U.S. Geological Survey Professional Paper 1270.

Swanson, V. E. (1961). *Geology and Geochemistry of Uranium in Marine Black Shales: A Review.* U.S. Geological Survey Professional Paper 356-C.

The Sunday Times (London). (2001). *Uranium Shortages Poses a Threat.* Available at: http://www.thetimes.co.uk [Accessed 10 January 2018].
World Nuclear Association. (2017). *Thorium.* Available at: http://www.world-nuclear.org [Accessed 20 January 2020].
World Nuclear Association. (2019). *World Uranium Mining Production.* Available at: http://www.world-nuclear.org [Accessed 20 January 2020].

Chapter 4

Low-Level Radioactive Wastes in the United States

"I began more carefully to look around me. A serious study of the soil was necessary to negate or confirm my hypothesis."

— Henry Lawson in *Journey to the Center of the Earth* by Jules Verne

4.1 Introduction

Low-level radioactive waste (LLRW) is a broad category of solids, liquids and gases that can contain several different radionuclides that are by-products from several sources. Nuclear power plants generate the majority of LLRW in the US (Contreras, 1992). In terms of energy production, these by-products result from the routine operations of nuclear power plants (Table 4.1), the decommissioning and demolition of former power plants and research reactors, and nuclear fuel fabrication and reprocessing facilities. The US Department of Energy (US DOE) produced LLRW from creating nuclear weapons, energy research, and site cleanup and decommission activities.

The specific radionuclides in LLRW from nuclear power plants include fission products, actinides and decay products, and neutron activation products. Fission products and actinides are produced by neutron absorption by nuclear fuel. Activation products are those created by neutron capture by other materials, such as structural components of the nuclear reactor and coolant system. Examples of activation products are given in Table 4.2.

Nuclear medicine generates LLRW. The radionuclides used in medicine, however, have relatively short half-lives (Table 4.3).

Table 4.1. Physical description of solid and liquid LLRWs generated by nuclear power plants (Saling and Fentiman, 2002).

<table>
<tr><td>Liquids</td><td>Chemical waste solutions, decontamination fluids, liquid-scintillation fluids, and contaminated oil.</td></tr>
<tr><td>Wet solids</td><td>Evaporator residues, spent ion-exchange media, spent filters.</td></tr>
<tr><td colspan="2">Solids (dry active wastes[a])</td></tr>
<tr><td>Compactible</td><td>Non-compactible</td></tr>
<tr><td>• Plastics: Coveralls, protective suits, gloves, hats, bags, bottles.
• Plastics: Coveralls, protective suits, gloves, hats, bags, bottles.
• Paper: Adsorbent paper, cartons.
• Absorbents: Vermiculite, bentonite.
• Cloth: Coveralls, laboratory coats, rags, gloves.
• Rubber: Boots, hoses, gloves.
• Wood: Construction lumber, plywood packing.
• Metals: Paint cans, aerosol cans.
• Glass: Bottles, laboratory glassware, faceplates.
• Filters: Air filters, respiratory canisters.</td><td>• Metals: Small tools, filter frames.
• Wood: Construction lumber, plywood packing.
• Conduit: Tubing, cable, wire, electrical fittings, pipes, valves.
• Concrete debris.
• Floor sweepings, contaminated soil.
• Glass: Lab glassware, instrument tubing.
• Lead: Shielding material.</td></tr>
</table>

[a]A dry active waste is the result of work conducted in an area that is potentially contaminated with radionuclides.

The US Nuclear Regulatory Commission (US NRC) concluded that LLRW with half-lives less than or equal to 120 days may be managed by decay-in-storage (DIS). When the radiation levels of the waste are indistinguishable from background levels, the waste may be disposed as ordinary trash or medical waste. While the wastes are in the storage facility, radiation doses must be as small as possible via ALARA (see Chapter 1). After the medical procedure, each of these short-lived radionuclides is eliminated from the body in urine. There has been a concern that body fluids containing medical radionuclides could pose a significant dose to sewage treatment workers, suggesting that the waste fluids should also be managed by DIS. Barquero *et al.* (2008) conducted a modeling study considering ^{67}Ga, ^{123}I, ^{131}I,

Table 4.2. Radiochemical properties of selected activation products (Audi *et al.*, 2003).

Activation product	Half-life	Decay mode[a]	Reaction
Tritium ($^{3}H_{1}$)	12.32 years	β^-	$^{3}H_{1} \rightarrow {}^{3}He_{2} + e^-$
Beryllium-10 ($^{10}Be_{4}$)	1.50×10^{6} years	β^-	$^{10}Be_{4} \rightarrow {}^{10}B_{5} + e^-$
Carbon-14 ($^{14}C_{6}$)	5,700 years	β^-	$^{14}C_{6} \rightarrow {}^{14}Na_{7} + e^-$
Sodium-24 ($^{24}Na_{11}$)	14.96 h	β^-	$^{24}Na_{11} \rightarrow {}^{24}Mg_{12} + e^-$
Sulfur-35 ($^{35}S_{16}$)	87.5 days	β^-	$^{35}S_{16} \rightarrow {}^{35}Cl_{17} + e^-$
Chlorine-36 ($^{36}Cl_{17}$)	301,000 years	β^-, EC	$^{36}Cl_{17} \rightarrow {}^{36}Ar_{18} + e^-$ $^{36}Cl_{17} + e^- \rightarrow {}^{36}S_{16}$
Iron-55 ($^{55}Fe_{26}$)	2.74 years	EC	$^{55}Fe_{26} + e^- \rightarrow {}^{55}Mn_{25}$
Nickel-59 ($^{59}Ni_{28}$)	76,000 years	EC	$^{59}Ni_{28} + e^- \rightarrow {}^{59}Co_{27}$
Cobalt-60 ($^{60}Co_{27}$)	5.27 years	β^-	$^{60}Co_{27} \rightarrow {}^{60}Ni_{28} + e^- + \gamma$
Nickel-63 ($^{63}Ni_{28}$)	100.1 years	β^-	$^{63}Ni_{28} \rightarrow {}^{63}Cu_{29} + e^-$
Molybdenum-93 ($^{93}Mo_{42}$)	4,120 years	EC	$^{93}Mo_{42} + e^- \rightarrow {}^{93}Nb_{41}$
Niobium-94 ($^{94}Nb_{41}$)	2.0×10^{4} years	β^-	$^{94}Nb_{41} \rightarrow {}^{94}Mo_{42} + e^-$
Technetium-99 ($^{99}Tc_{43}$)	211,100 years	β^-	$^{99}Tc_{43} \rightarrow {}^{99}Ru_{44} + e^-$
Cadmium-113m ($^{113m}Cd_{48}$)	14.1 years	β^-	$^{113m}Cd_{48} \rightarrow {}^{113}In_{49} + e^-$
Lead-205 ($^{205}Pb_{82}$)	1.53×10^{7} years	EC	$^{205}Pb_{82} + e^- \rightarrow {}^{205}Tl_{81}$
Polonium-210 ($^{210}Po_{84}$)	138.38 days	α	$^{210}Po_{84} \rightarrow {}^{206}Pb_{82} + \alpha$

[a]See Chapter 1 for definitions of radioactive decay modes.

^{111}In, ^{99m}Tc, and ^{201}Tl, and concluded that the accumulated dose of 29 μSv/year (2.9 mrem/year) to sewage workers was not a health concern.

There are more than 4,000 industrial generators of LLRW in the US (Contreras, 1992) in connection with the production of radioactive chemicals for agriculture, environmental, pharmaceutical,

Table 4.3. Radionuclides used in medicine (Audi *et al.*, 2003).

Isotope	Half-life	Decay mode	Reaction
Imaging			
Fluorine-18 ($^{18}F_9$)	109.8 min	β^+	$^{18}F_9 \rightarrow {}^{18}O_8 + e^+$
Gallium-67 ($^{67}Ga_{31}$)	3.26 days	EC	$^{67}Ga_{31} + e^- \rightarrow {}^{67}Zn_{30}$
Krypton-81m ($^{81m}Kr_{36}$)	13.1 s	IT	$^{81m}Kr_{36} \rightarrow {}^{81}Kr_{36} + \gamma$
Rubidium-82 ($^{82}Rb_{37}$)	1.27 min	β^+	$^{82}Rb_{37} \rightarrow {}^{82}Kr_{36} + e^+$
Technetium-99m ($^{99m}Tc_{43}$)	6.02 h	IT	$^{99m}Tc_{93} \rightarrow {}^{99}Tc_{93} + \gamma$
Indium-111 ($^{111}In_{49}$)	2.8 days	EC	$^{111}In_{49} + e^- \rightarrow {}^{111}Cd_{48}$
Iodine-123 ($^{123}I_{53}$)	13.2 h	EC	$^{123}I_{53} + e^- \rightarrow {}^{123}Te_{52}$
Xenon-133 ($^{133}Xe_{54}$)	5.25 days	β^-	$^{133}Xe_{54} \rightarrow {}^{133}Cs_{55} + e^-$
Thallium-201 ($^{201}Tl_{81}$)	72.9 h	EC	$^{201}Tl_{81} + e^- \rightarrow {}^{201}Hg_{80}$
Therapy			
Strontium-89 ($^{89}Sr_{38}$)	50.5 days	β^-	$^{89}Sr_{38} \rightarrow {}^{89}Y_{39} + e^-$
Yttrium-90 ($^{90}Y_{39}$)	64.0 h	β^-	$^{90}Y_{39} \rightarrow {}^{90}Zr_{40} + e^-$
Iodine-125 ($^{125}I_{53}$)	59.4 days	EC	$^{125}I_{53} + e^- \rightarrow {}^{125}Te_{52} + \gamma$
Iodine-131 ($^{131}I_{53}$)	8.02 days	β^-	$^{131}I_{53} \rightarrow {}^{131}Xe_{54} + e^- + \gamma$

and biomedical uses. Consumer products such as smoke detectors, enamel glazes, and illuminated signs can generate LLRW. Universities and hospitals conducting biomedical research also generate LLRW, including wastes from research projects and instruments, and carcasses of animals treated with radioactive materials used in medical or pharmaceutical research.

4.2 Protection from Low-Level Radioactive Wastes

Based on studies conducted by the International Commission on Radiological Protection (ICRP, 1977, 1979), the US NRC proposed a whole-body dose equivalent of 5 mSv/year (500 mrem/year) for the "critical group" which was the group of people who were most likely to receive the largest radiation dose during some given activity (Adams and Rogers, 1978). The US NRC developed and promulgated specific requirements for LLRW disposal facilities (US NRC, 1981). The rule is known as 10 (Code of Federal Regulations (CFR) Part

61 Licensing Requirements for Land Disposal of Radioactive Waste or in this textbook as "Part 61." It is based on passive rather than active controls to minimize and retard the release of radionuclides into the environment. The intent of Part 61 is to provide protection during the routine disposal operations as well as long-term protection after the facility has been closed. This protection includes accidental exposures caused by an inadvertent human intruder who is not aware of the presence of LLRW at the site. It was the intent of the US NRC that future generations should be provided the same level of protection as the general population during the active operation of the disposal site. The rule was not applicable to the then-existing LLRW sites. The requirements not only apply to shallow-land disposal, but to below-ground vaults, earth-mounded concrete bunkers, and augured holes (Ryan *et al.*, 2007).

4.3 Low-Level Radioactive Waste Classification

Based on the modeling results and the input of the scientific and engineering community, the US NRC developed a LLRW classification system. The basis of this system is the physical stability of the waste container and the concentrations of the radionuclides. The classes of LLRW that are acceptable for shallow-land disposal are simply called Class A, B, and C (Table 4.4). Class A is the least radioactive, and therefore has no special stability requirements. Therefore, it must be separated from other LLRW because of the potential for the waste containers to become compacted with time because of decomposition and overburden pressures. Compaction could result in subsidence of the ground surface above the disposal cells. Class B LLRW requires physical stability of the waste containers and contains larger concentrations of radionuclides. Although Class C is acceptable for shallow-land disposal, it requires deeper burial and some type of engineered barrier to protect the inadvertent intruder after the institutional controls lapse. LLRW that exceeds Class C criteria are regarded as unacceptable for shallow-land disposal. Such wastes, the most radioactive of all LLRW are called Greater Than Class C (GTCC). The reader may find it informative to

Table 4.4. Summary of the LLRW classes.

Class A

Smallest level: Any substance containing any radionuclide is Class A. There is currently no "below regulatory-concern" level.

- Specific activity: Near background to 700 Ci/m^3 (25.9 TBq/m^3)
- No special stability requirements such as solidification of waste packages. Cannot use cardboard boxes, however.
- Not a threat to the inadvertent intruder beyond 100 years: exposure to the intruder is less than 500 mrem/year (5 mSv/year).

Class B

Middle level

- Specific activity: 0.04–700 Ci/m^3 (1,480 MBq to 25.9 TBq)
- The waste container must be designed to be stable for 300 years (solidification of stable waste packages).
- Not a threat to the inadvertent intruder beyond 100 years; exposure to the intruder is less than 500 mrem/year (5 mSv/year).
- Must be segregated from Class A wastes at the site.

Class C

Largest level acceptable for shallow land disposal

- Specific activity: 44 to 7,000 Ci/m^3 (1.63–259 TBq).
- The waste container must be designed to be stable for 300 years (solidification of stable waste packages).
- Requires some type of engineered barrier.
- Will not endanger inadvertent intruder beyond 500 years; exposure to the intruder is less than 500 mrem/year (5 mSv/year).
- Must be segregated from Class A wastes at the site.

GTCC

Not acceptable for shallow-land disposal

- Poses a threat to the inadvertent intruder beyond 500 years.
- Must be disposed in a geological repository or an alternative proposed by US DOE and approved by US NRC. Examples: sealed sources commonly ^{137}Cs and ^{241}Am, activated metals, decommissioning wastes.

compare the US LLRW classification system with those in other countries in Chapter 10, and the radioactive waste classification system provided by the International Atomic Energy Agency in Appendix B.

The US NRC classifies LLRW by using two tables: one for long-lived radionuclides (Table 4.5), and one for short-lived isotopes (Table 4.6). For regulatory purposes, the criteria for defining "short-lived" from "long-lived" is 100 years. The group of alpha-emitting transuranics includes ^{237}Np, ^{238,239,240,242,244}Pu, ^{241,242m,243}Am, ^{243,244,245,246,247,248}Cm, ^{247}Bk, ^{249,250,251}Cf. Most wastes contain mixtures of radionuclides. To classify a waste with more than one radionuclide, the Sum-of-the-Fractions Rule is used. For example, suppose that a waste contains strontium-90 at a concentration of 50 Ci/m^3 (1.85 TBq/m^3) and 22 Ci/m^3 (0.81 TBq/m^3) of cesium-137. According to Table 4.6, both are greater than the criteria for Class

Table 4.5. The US LLRW classification table for radionuclides with a half-life longer than 100 years.

Radionuclide	Class A	Class C
Carbon-14 (^{14}C)	≤0.8 Ci/m^3 (≤0.03 TBq/m^3)	>0.8 Ci/m^3 (>0.03 TBq/m^3)
^{14}C in activated metal	≤8.0 Ci/m^3 (≤0.30 TBq/m^3)	>8.0 Ci/m^3 (>0.30 TBq/m^3)
Nickel-59 (^{59}Ni) in activated metal	≤2.20 Ci/m^3 (≤0.08 TBq/m^3)	>2.20 Ci/m^3 (>0.08 TBq/m^3)
Niobium-94 (^{94}Nb) in activated metal	≤0.02 Ci/m^3 (≤0.0007 TBq/m^3)	>0.02 Ci/m^3 (>0.0007 TBq/m^3)
Technetium-99 (^{99}Tc)	≤0.30 Ci/m^3 (≤0.01 TBq/m^3)	>0.30 Ci/m^3 (>0.01 TBq/m^3)
Iodine-129 (^{129}I)	≤0.008 Ci/m^3 (≤0.0003 TBq/m^3)	>0.008 Ci/m^3 (>0.0003 TBq/m^3)
Plutonium-241 (^{241}Pu)	≤0.35 pCi/g (≤0.013 Bq/g)	>0.35 pCi/g (>0.013 Bq/g)
Curium-242 (^{242}Cm)	≤2 pCi/g (≤0.074 Bq/g)	>2 pCi/g (>0.074 Bq/g)
Alpha-emitting transuranics with a half-life longer than 5 years	≤10 nCi/g (≤370 Bq/g)	>10 nCi/g (>370 Bq/g)

Table 4.6. The US LLRW classification table for radionuclides with a half-life longer than 100 years.

Radionuclide	Class A	Class B	Class C
Tritium (^{3}H)	40 Ci/m^3 (1.48 TBq/m^3)	No limit	No limit
Cobalt-60 (^{60}Co)	700 Ci/m^3 (25.9 TBq/m^3)	No limit	No limit
Nickel-63 (^{63}Ni) in activated metal	3.5 Ci/m^3 (0.13 TBq/m^3)	70 Ci/m^3 (2.59 TBq/m^3)	700 Ci/m^3 (25.9 TBq/m^3)
Strontium-90 (^{90}Sr)	0.04 Ci/m^3 (0.15 TBq/m^3)	150 Ci/m^3 (5.55 TBq/m^3)	7,000 Ci/m^3 (259 TBq/m^3)
Cesium-137 (^{137}Cs)	1 Ci/m^3 (0.04 TBq/m^3)	44 Ci/m^3 (1.63 TBq/m^3)	4,600 Ci/m^3 (170 TBq/m^3)
Total of all radionuclides with a half-life that is shorter than 5 years	700 Ci/m^3 (25.9 TBq/m^3)	No limit	No limit

A. Is the mixture Class B or Class C? To determine the class of the mixture, concentration of each radionuclide is divided by limit for the next class, in this case Class B, and the results are then added:

$$\begin{aligned}\text{For } ^{90}\text{Sr:}\ & (50\,\text{Ci/m}^3)/(150\,\text{Ci/m}^3) = 0.33. \text{ In SI units,}\\ & (1.85\,\text{TBq/m}^3)/(5.55\,\text{TBq/m}^3) = 0.33\\ \text{and } ^{137}\text{Cs:}\ & (22\,\text{Ci/m}^3)/44\,\text{Ci/m}^3) = 0.50. \text{ In SI units,}\\ & (0.81\,\text{TBq/m}^3)/(1.63\,\text{TBq/m}^3) = 0.50\end{aligned}$$

Adding 0.33 + 0.50 yields 0.83. Because the result is <1.0, the mixture meets the criteria to be classified as Class B.

4.4 Additional Radioactive Materials and Wastes

In addition to the system used by the US NRC to classify LLRW (Table 4.4), the reader may encounter other classification terms that have been in the US during the management and disposal of radioactive sources. Some of the terms are used informally and are applied to radioactive materials that are not regulated by the US

NRC. The information in this section was taken from US EPA (2019) and US NRC (2019).

4.4.1 *Exempt*

Radioactive materials that are consumer products that do not contain a sufficient amount of a radionuclide(s) to pose a significant radiological risk to the public are defined as Exempt. Examples include electronic components that contain cesium-137, cobalt-60, radium-226, krypton-85, and others, gunsights containing tritium, incandescent gas mantles that contain thorium-232, and smoke detectors that contain americium-241.

4.4.2 *Source Material*

The US NRC defines "source material" as material that contains thorium or uranium provided that the uranium has not been enriched with respect to uranium-235. Depleted uranium is also classified as a Source Material.

4.4.3 *Special Nuclear Material*

The US NRC defines this material as plutonium, uranium-233, and any uranium that is enriched with ^{233}U or ^{235}U. This classification does not include Source Material.

4.4.4 *Low-Level Radioactive Waste*

The US DOE does not differentiate low-level wastes using the system of the US NRC. US DOE simply defines low-level wastes as "that is not high-level radioactive waste, spent nuclear fuel, transuranic waste, byproduct material or naturally occurring radioactive material" US DOE (1999). Byproduct material is defined as "tailings or wastes produced by the extraction or concentration of uranium or thorium from any ore processed primarily for its source material content" or any discrete source of radium-226.

4.4.5 *Mixed Wastes*

This type of waste is a mixture of radioactive material and hazardous components — hazardous as defined by the US Resource Conservation and Recovery Act (RCRA). The US EPA has defined specific criteria that are applied to solids, liquids, and gases to define the degree of hazard. The US EPA created standardized test protocols to measure ignitability, corrosivity, reactivity, and toxicity. The US NRC and the US DOE regulate the radioactive portion of a mixed waste, and the US EPA regulates the hazardous-waste portion under the authority of RCRA. Mixed Wastes are also known as Mixed Radiological and Hazardous Waste. Most commercially-generated mixed waste is classified as Low-Level Mixed Waste (abbreviated as LLMW or MLLW). The US DOE produces three types of mixed wastes:

Low-Level Mixed Waste (LLMW)
High-Level Mixed Waste (HLMW)
Mixed Transuranic Waste (MTRU)

4.4.6 *Low-Activity Radioactive Waste (LARW)*

The US EPA informally defines LARW as any radioactive material that contains relatively small amounts of radionuclides. "Low activity" is not equivalent to the US NRC classification of "low level." (Table 4.4). LARW is regarded as a "concept" whereas LLRW is a regulatory definition. Radioactive wastes from several sources have the potential to be defined as low-activity wastes. These sources include Naturally Occurring and Accelerator-Produced Material (NARM) and Naturally Occurring Radioactive Material (NORM). Both sources refer to a broad category of materials that contain radionuclides in which human activities may increase the potential of human exposure to the ionizing radiation from these materials (WNA, 2019). The term NORM is used to differentiate material containing naturally-occurring radionuclides from anthropogenic radionuclides such as those produced by nuclear power and medicine. The radionuclides in NORM are often uranium and its decay chain, thorium, and potassium-40.

When the radionuclides in NORM become concentrated by some anthropogenic activity or process, the resulting product is called Technologically Enhanced Naturally Occurring Radioactive Material (TENORM). The difference between a NORM and a TENORM is a subject of debate because both materials can result from the application of *some* type of technology that increases the risk of human exposure — such as mineral mining. A classic example of NORM is coal. Coal contains naturally occurring radioactive elements — most importantly uranium-238, thorium-232, radium-226, and potassium-40 (EPRI, 2014). When coal is combusted, these radionuclide become concentrated in coal ash which is derived from the noncombustible mineral matter in coal. Coal ash is a TENORM. Oil and natural gas production also produce NORM in the form of a wastewater that can contain radium-226 and -228 (WNA, 2019). Sand deposits that are mined for economic minerals may be NORM because of the presence of thorium-232. Phosphorite is mined and extracted to make phosphorus fertilizer which is NORM. The process mobilizes radium-226, thorium-232, and uranium-238 that occur in the sedimentary source rock. Building materials such as granite, concrete, and clay bricks are all NORM because they contain radium-226, thorium-232, and potassium-40.

4.4.7 *Residual Radioactive Material (RRM)*

RRM is the US DOE-US NRC term for mill tailings resulting from the processing of ores for the extraction of uranium and other valuable constituents in the ores. The US NRC does not consider mill tailings to be LLRW.

4.5 Potential Revisions of 10 CFR Part 61

The US NRC finalized 10 CFR Part 61 in 1982, and it is the basis for LLRW classification and management in the US. Beginning in about 2010, however, the US NRC began considering whether the current LLRW criteria should be revised. US DOE (2012) summarized that three issues suggest the need to revise Part 61:

1. Emergence of potential LLRW streams that were not considered during the initial Part 61 rulemaking such as depleted uranium and blended LLRW.[1]
2. US DOE's increasing use of commercial facilities for the disposal of defense-related LLRW.
3. International experience gained in managing LLRW and intermediate-level radioactive wastes[2] that did not exist when Part 61 was promulgated.

The process is on-going.

4.6 Early LLRW Disposal

During the early years of the domestic nuclear energy industry, the Atomic Energy Commission (AEC) disposed of both commercial and Federal LLRW. The wastes were often placed in 55-gallon steel drums, mixed with cement or concrete, and then disposed in the oceans at depths greater than 1,830 m (Ryan *et al.*, 2007). There were five major disposal areas in the Pacific Ocean, one in the Gulf of Mexico, and 11 in the Atlantic Ocean. For example, it has been estimated that, between 1951 and 1967, about 34,203 containers holding 2,941 TBq (79,483 Ci) at the time of packaging were disposed in the Atlantic Ocean (Ryan *et al.*, 2007) (see Chapter 5). Also, during the 1950s, shallow land burial was used at Federal facilities. Cardboard boxes and open drums containing LLRW were dumped into unlined ditches on the Hanford Site (Fig. 4.1).

By about 1970, ocean disposal of LLRW by the US ended because of adverse public reaction to the practice, and because ocean disposal became more expensive than shallow land disposal. In 1972, the US joined the London Convention which is an international agreement to

[1]Waste blending means mixing wastes of different concentrations prior to disposal. For example, Class B or C waste could be mixed with a sufficient amount of Class A material such that the overall mixture is Class A. Current NRC regulations to not address this practice.

[2]International LLRW classification and definitions are given in Appendix B.

Figure 4.1. LLRW dumping at the Hanford Site in the 1950s (Gephardt, 2010). Used with permission of Elsevier.

ban ocean disposal of radioactive wastes. The AEC had endorsed the idea of shallow land burial in commercial disposal sites. During the interim before a commercial facility was built, LLRW was disposed at about 16 federally owned sites.

4.6.1 *Beatty, Nevada*

In 1962, the Beatty Low Level Radioactive Waste Site became the first commercially operated facility for LLRW to be license by the US AEC. Located in the Amargo Desert in Nevada, the site geology consisted of alluvial sand, silt, clay, and gravel to depth of about 183 m (600 feet) (Fig. 4.2). The regional groundwater table occurs at a depth of 79.3–100.6 m (260–330 feet) (INEL, 1994). The climate at this site is arid, with an annual rainfall of 6.4–12.7 cm (2.5–5 inches). The disposal area consists of 22 trenches which vary is depth from 1.8 to 15.2 m (6–50 feet). When the trenches were filled, a trench cover was mounded over the wastes.

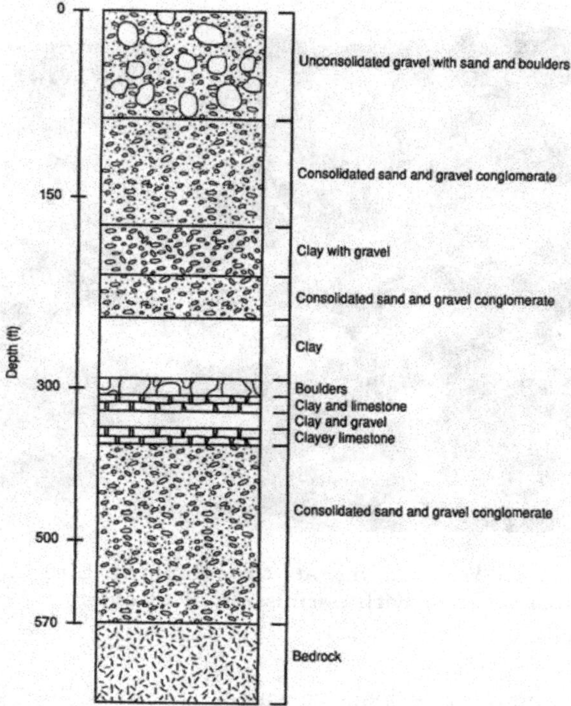

Figure 4.2. Geological column of the Beatty LLRW Site (INEL, 1994). The layers of sand, gravel, and boulders would likely not serve as a barrier to leachate movement from the disposal trenches. Although the geology of the area is not well suitable for a LLRW site, groundwater occurs at a relatively deep depth, and the site is located in an arid climate which limits the production of leachate.

From 1962 to 1992, 133,930 m^3 (4,729,690 ft^3) of LLRW were buried which represented about 23,680 TBq (640,530 Ci) of activity. According to INEL (1994), there were no "significant environmental problems" with respect to shallow groundwater contamination at the Beatty facility. Wilshire and Friedman (1999), however, presented field data that tritium had migrated off-site. Tritium was detected in water samples collected from the unsaturated zone and from shallow groundwater monitoring wells. Gross alpha and gross beta levels were also greater than background levels, but the specific radionuclides generating the activity were not identified. Wilshire and Friedman reported that liquid wastes had been placed in the trenches from about 1966 to 1975, and that it was not usual for trenches to be kept open for periods of as long as 141 months before closure, allowing even the sparse precipitation to enter the trenches. The site closed in 1992, in part, because it had a history of facility management and waste transportation mistakes which resulted in poor public relations. Also, it was anticipated that a new LLRW facility was to be developed in Colorado for the Rocky Mountain Compact. The new site was not built.

4.6.2 *Maxey Flats, Kentucky*

In 1963, the Maxey Flats Disposal Site opened under a lease between the State of Kentucky and Nuclear Engineering Company, Inc. The site geology consisted of thin and relatively thicker layers of shale, sandstone, and siltstone to a depth of about 39.6 m (130 feet). There is a continuous groundwater table at a depth of about 9.1–15.2 m (30–50 feet) (INEL, 1994). The facility is in a humid region with annual rainfall of 117 cm (46 inches). The disposal area consists of 52 trenches, smaller pits, and vertical disposal wells (Fig. 4.3). The depth of the trenches varies from 2.7 to 9.1 m (9–30 feet), and they were sloped to facilitate water collection and removal. The disposal wells are 4.6 m (15 feet deep) and capped with concrete. They were used to dispose of high-activity gamma sources. The pits are between 1.5 and 4.6 m (5 and 15 feet) deep.

Maxey Flats became the largest commercial repository for LLRW in the US. Between 1963 and 1977, 135,265 m^3 (4,776,840 ft^3) of wastes were buried which represented about 88,826 TBq (2,400,690 Ci) of activity. During its operation, Maxey Flats was plagued with several siting and operational practices that, in part, became the rationale for the modern waste-disposal criteria. The waste form was likely one of the most significant factors that lead to the demise of Maxey Flats. As was the practice at the time, neither waste minimization nor volume reduction was required of LLRW prior to shallow land disposal. The waste packages buried at Maxey

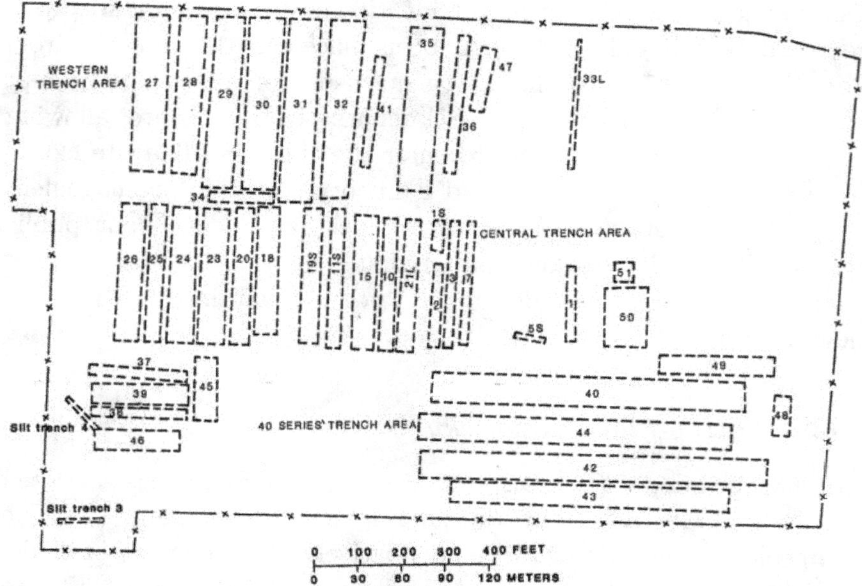

Figure 4.3. Plan view of Maxey Flats Disposal Site showing the disposal trenches. The narrow and slit trenches were used for the more highly radioactive wastes to minimize occupational exposure (Wilson and Lyons, 1991).

Flats were believed to be a mixture of easily degradable cardboard and fiberboard boxes mixed with wooden boxes and 55-gallon drums. There was also little or no attempt to separate the waste packages by the relative stability of neither the waste containers nor their chemical content. Also, the waste containers were not compacted prior to placement. Because of the random placement of the containers after burial, there were initially void spaces between the containers. As the waste containers decomposed, the backfill soil settled to fill the void spaces, resulting in the subsidence and eventually failure of the trench covers to isolate the wastes from the environment.

The failure of the covers was accelerated by excessive moisture. The permeability of the weathered bedrock (about 0.02 cm/day)

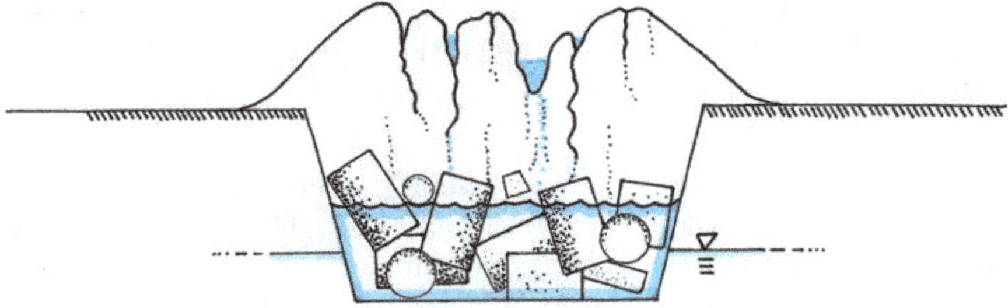

Figure 4.4. Diagram of the "bathtub effect." The waste containers were randomly placed in the disposal trench followed by settlement which compromises the trench cover, allowing precipitation to enter the trench. The zone of saturation is close to the surface providing groundwater to the trench. The trench fills with water/leachate and can overflow. The figure is not to scale.

was regarded as a positive attribute in site selection; the slow flow rate of shallow groundwater would result in effective containment of the radionuclides (Wheeler and Dragonette, 1981). However, because of the humid climate, water accumulated in many of the disposal trenches resulting in the "bathtub effect" (Fig. 4.4). The excessive moisture further accelerated the decomposition of the waste containers, and ultimately the failure of the trench covers which allowed more infiltration into the trench.

In 1973, the State of Kentucky required Nuclear Engineering Company, Inc. to begin a water management program to collect and pump contaminated water from the trenches into above-ground storage tanks. The program was poorly conducted and resulted in surface spills and the direct discharge of contaminated water to the surface (INEL, 1994). In 1977, it was discovered that trench leachate was migrating laterally along a relatively thin layer of siltstone that was widespread through out the site at a depth of about 7.6 m (25 feet) below ground surface. The leachate had migrated into a newly dug trench. The Commonwealth of Kentucky ordered the site closed. In the early 1980s, tritium (a radioactive isotope of hydrogen) was detected in surface water, groundwater, and in vegetation in the west side of the facility. In 1986, Maxey Flats was designated by the

US EPA as a Superfund Site. The US EPA notified 832 Potential Responsible Parties which included the University of Illinois at Urbana-Champaign. During the remediation period that followed, an effort was made to extract and treat trench leachate. Then a 45-mil (1.14-mm thick) geomembrane liner was installed to cover the trench area to prevent the infiltration of water. The site is currently managed by the Environmental and Public Protection Cabinet of the Commonwealth of Kentucky. The site is considered unusable for any future purpose and will have to be monitored and maintained. A 100-year institutional-control period is in progress. The reader will find recent information about the status of the Maxey Flats Disposal Site in US EPA (2017).

4.6.3 *West Valley, New York*

The West Valley Nuclear Waste Site opened as a commercial facility in 1963 by Nuclear Fuel Services, Inc. The site consists of 14 parallel trenches that are about 6.1 m (20 feet) deep. The site geology consists of 3.0–6.1 m (10–20 feet) of weathered glacial till over 45.7–91.4 m (150–300 feet) of unweathered till composed of layers of sand, gravel, silt, and clay. The bedrock at the facility is shale. When a section of a trench was filled, it was covered with 1.2–2.4 m (4–8 feet) of soil. During its operational period, 69,860 m^3 (2,467,160 ft^3) of LLRW were buried which equated to 46,705 TBq (1,262,300 Ci) of activity (INEL, 1994).

Like Maxey Flats, West Valley was plagued by the "bathtub effect." In 1975, two trenches became filled with leachate, and seeped through covers. The LLRW disposal facility was then closed. Since that time, leachate has been collected and pumped into above-ground storage tanks. A combination of plastic trench covers, re-compacted soil covers, and concrete barriers have been installed at various stages to reduce the infiltration of precipitation into the trenches. Because the LLRW site is in the middle of the Nuclear Services Center, it is difficult to assess what impact the LLRW facility had on groundwater quality. The future of now closed LLRW facility is ultimately linked to the future of the West Valley Demonstration Project.

4.6.4 *Richland, Washington*

The US Ecology Richland Washington Facility for LLRW opened in 1965 as a commercial business. The facility is unique because it is located on Federal land within the 1,450-km^2 (560-mile^2) US DOE Hanford Site, which is located on a semi-arid river plain along the Columbia River. The annual precipitation is about 16.0 cm (6.3 inches) (INEL, 1994). The site geology consists of a mixture of sand, silt, and gravel that is about 75 m (200 feet) thick which overlies a thicker sequence of unconsolidated material. The depth to the water table is about 75 m (245 feet).

The Richland Washington Facility has a long, complicated history. It was first operated by California Nuclear, Inc. and began accepting LLRW and chemical wastes. In 1980, packing requirements became more stringent, and cardboard and fiberboard packaging were prohibited. Wooded boxes were prohibited in 1987. Because of its proximity to Hanford, it is difficult to assess the environmental impacts *specific* to LLRW at Richland. Gephart (2010) noted that in the late 1990s, there were about 75 LLRW burial grounds at Hanford, holding about 700,000 m^3 of waste, and that possibly 37 PBq (1 million Ci) of activity remain in the soil and groundwater beneath Hanford.

The Richland Washington Facility is still open. It is currently accepting Class A, B, and C LLRW from 11 Compact States (Northwest and Rocky Mountain) states: Alaska, Hawaii, Idaho, Montana, Oregon, Utah, Washington, Wyoming, Colorado, Nevada, and New Mexico. It accepts NORM, NARM, high-activity radium wastes, smoke detectors, and other Exempt Wastes from all 50 states. From 2005 to 2017, the Richland Washington Facility disposed only 1.25% of the total volume of LLRW disposed in the US (Table 4.7), and about 5.6% of the total amount of radioactivity associated with the wastes disposed during the same period.

4.6.5 *Sheffield, Illinois*

The Sheffield LLRW Disposal Facility opened in 1968, and was first operated by California Nuclear, Inc. (Fig. 4.5). The site geology

Table 4.7. Compilation of the volume and activity of LLRWs disposed at the four currently operating disposal facilities in the US.

Disposal facility	Volume	Activity
Texas Compact Waste Facility (Texas)	3,231 m^3 114,100 ft^3	15,641 TBq 422,718 Ci
Barnwell Disposal Facility (South Carolina)	7,341 m^3 259,235 ft^3	102,645 TBq 2,774,177 Ci
Clive Disposal Facility (Utah)	878,046 m^3 31,007,890 ft^3	3,696 TBq 99,878 Ci
US Ecology Washington (Washington State)	11,275 m^3 398,172 ft^3	7,186 TBq 194,212 Ci
Total of the four facilities	899,892 m^3 31,779,397 ft^3	129,166 TBq 3,490,983 Ci

Note: The time period is from 2005 to 2017 except for the Texas facility which accepted wastes from 2012 to 2017. These disposal amounts were based on data provided by the US NRC. Available at: https://www.nrc.gov/waste/llw-disposal/licensing/statistics.html#2007 [Accessed 26 January 2020].

Figure 4.5. Waste container placement at the Sheffield LLRW Disposal Facility. As typical in that era, there was little effort to stack the containers to minimize subsidence (photograph taken by the author).

consists of sand over lake deposits of silt which overlie pebble-rich sand to a depth of about 61.0 m (200 feet). Below that, the bedrock is shale (INEL, 1994). The climate is humid, with a mean annual precipitation of about 88.9 cm (35 inches). The depth of the saturated zone is about 7.6 m (25 feet) (IEMA, 2009).

The site consists of 8.1 ha (20) acres of 21 unlined trenches that are 6.1–7.6 m (20–25 feet) deep. The site was used for disposal until 1978 by which time 88,324 m^3 (3,119,138 feet^3) of waste had been buried which equaled about 2,228 TBq (60,206 Ci) of activity (INEL, 1994; IEMA, 2009). The disposed waste included 57 kg of Special Nuclear Material.

In 1977, tritium was detected in on-site monitoring wells. It appeared that the tritium had migrated more rapidly than expected, based on the initial site characterization. Follow-up site investigations revealed that there were continuous, water-permeable sand layers that would allow leachate to migrate from the disposal trenches. Because of this discovery, the US NRC decided that a newly constructed trench could not be used for waste disposal. Because all of the other trenches had been filled to capacity, the waste facility was, by default "full" and ceased accepting LLRW in 1978. Another possible problem was that the trench covers were not sufficiently compacted to reduce infiltration of precipitation and cap subsidence (IEMA, 2009).

In 1981, tritium was detected off-site (Fig. 4.6). A 1.4-m (4.5-foot) layer of compacted clay was placed over the trench area to reduce infiltration. IEMA has implemented a monitoring program in and around the Sheffield site that includes, surface water, off-site public and private wells, and on-site groundwater wells. The Sheffield site is owned by the State of Illinois, while a buffer zone around the site is owned and monitored by US Ecology until 2038 (IEMA, 2009).

4.6.6 *Barnwell, South Carolina*

The Barnwell Disposal Facility began disposing LLRW in 1971, and is operated by Chem-Nuclear Systems, Inc. The site geology consists of layers of sand and gravel to a depth of more than 152 m (500

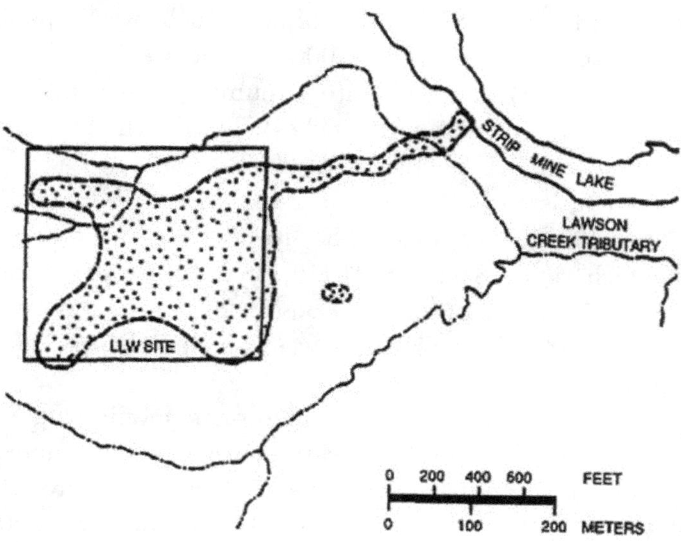

Figure 4.6. Movement of tritium off-site from the Sheffield LLRW Disposal Facility (modified from Kelly, 1987).

feet). The regional water table is at a depth of about 9.1–18.3 m (30–60 feet) (INEL, 1994). The climate is humid with a mean annual precipitation of 119 cm (47 inches). The Barnwell Disposal Facility occupies 95.1 ha (235 acres) (BLWM, 2007). Barnwell can currently accept Class A, B, and C wastes for shallow land burial, but only from three member states of the Atlantic Compact (Connecticut, New Jersey, and South Carolina). Prior to 2008, Barnwell could accept all waste classes from non-compact States. In 2001, South Carolina enacted legislation to terminate access to non-compact states in 2008. The action affected 36 US States who had been sending Class B and C wastes to Barnwell.

Class A trenches are about 305 m (100 feet) long, 91.4 m (300 feet) wide, and 9.1 m (30 feet) deep. Typical Class B and C waste trenches are 182.9 m (600 feet) long, 15.2 m (50 feet) wide, and 6.1 m (20 feet) deep (BLWM, 2007). Slit trenches are used for "high Class C" concentrations. Regardless of waste classification, the wastes are placed in concrete overpacks. Rectangular vaults that are

3.1 m × 3.1 m × 3.4 m (10 feet × 10 feet × 11 feet) tall are used for Class A wastes in metals boxes and drums. Cylindrical concrete vaults are used for Class B and C wastes. The use of vaults is thought to help stabilize the trenches against subsidence after the trenches are backfilled. The trenches are covered with a cap composed of a sand-drainage layer, plastic sheet, compacted clay-soil liner, and sandy topsoil all designed to reduce infiltration into the trench.

From 2005 to 2017, the Barnwell Disposal Facility disposed only 0.8% of the total volume of LLRW disposed in the US (Table 4.7). In contrast, however, about 79.5% of the total amount of radioactivity associated with the wastes disposed during the same period in the US reside at Barnwell. The disproportionately large amount of radioactivity resulted because of the Barnwell's ability to accept Class A, B, and C wastes.

4.6.7 *Clive, Utah*

The Energy*Solutions* Clive Disposal Facility is the largest (259 ha or 640 acre) commercial LLRW disposal site in the US (Fig. 4.7). The facility is located in Utah's western desert. The site geology is composed of layers of silt, gravel, and sand. Sediments deposited by Pleistocene-age Lake Bonneville dominate the surficial geology. Lake Bonneville was a large ice-age lake that covered much of northwestern Utah between about 25,000 and 15,000 years ago. Unconsolidated material in the Clive disposal area includes gravel, sand, and fine-grained sediment deposited during various stages of the Bonneville lake cycle, alluvium deposited prior to and during lake advance, and alluvium deposited after lake retreat (Black *et al.*, 1999). The climate is arid with an annual precipitation of 20.3 cm (8 inches) per year (Ledoux and Cade, 2002).

The facility is located in the West Desert Hazardous Industry Area which includes two hazardous waste incinerators, and a hazardous-waste landfill. The site was initially used to dispose of uranium mill tailings in Denver. The Clive Facility is licensed to accept Class A LLRW, LARW, NORM, and MLLW. Class A wastes, LARW, and the Class A-mixed wastes are placed into separate

Figure 4.7. The Energy*Solutions* Clive Facility. In addition to showing the location of the disposal area of Class A LLRW, there are areas for uranium mill tailings from Salt Lake City Vitro Chemical, Byproduct Material (11e.(2)), Low Activity Radioactive Waste (LARW), and Class A-Mixed LLRW. CWF refers to Containerized Waste Facility (Energy*Solutions*, 2015). Used with permission by Energy*Solutions*.

disposal cells. The soil at the bottom of each cell is compacted to have a permeability of 10^{-6} cm/s (Ledoux and Cade, 2002). Each cell is then covered with soil to form a mostly above-grade embankment. A cover made of soil and gravel protects the embankment from erosion and plant growth. The cells are surrounded by a network of groundwater monitoring wells. The air, vegetation, and soils on and in close proximity to the site are also monitored.

From 2005 to 2017, the Clive Disposal Facility disposed 97.6% of the total volume of LLRW disposed in the US (Table 4.7). In contrast, however, only about 2.9% of the total amount of radioactivity associated with the wastes disposed during the same period in the US reside at Clive. The disproportionately small amount of radioactivity resulted because of the Facility's ability to accept only Class A wastes.

4.7 Low-Level Radioactive Wastes Compacts

The closure of LLRW disposal sites resulted in an increase in the volumes of wastes arriving at Richland Washington, Barnwell, and the Beatty Low Level Radioactive Waste Site. The governors of the US States of Washington, Nevada, and South Carolina wanted the rest of the country to "share the burden of LLRW disposal." In response, Congress passed the Low-Level Radioactive Waste Policy Act (LLRWPA) in 1980 (amended in 1985) to promote the siting and construction of new regional LLRW disposal facilities. Most States joined together to build regional facilities by forming organizations called compacts (Table 4.8). There had been, however, few attempts

Table 4.8. LLRWs compacts.

Northwest	**Midwest**	**Central Midwest**
Alaska	Indiana	Illinois
Hawaii	Iowa	Kentucky
Idaho	Minnesota	
Montana	Missouri	**Appalachian**
Oregon	Ohio	Delaware
Utah	Wisconsin	Maryland
Washington		Pennsylvania
Wyoming	**Central**	West Virginia
	Arkansas	
Rocky Mountain	Kansas	**Atlantic**
Colorado	Louisiana	Connecticut
Nevada	Oklahoma	New Jersey
New Mexico		South Carolina
	Texas	
Southwestern	Texas	**Southeast**
Arizona	Vermont	Alabama
California		Florida
North Dakota	**Unaffiliated States**	Georgia
South Dakota	Maine	Mississippi
	Massachusetts	Tennessee
	Michigan	Virginia
	Nebraska	
	New Hampshire	
	New York	
	North Carolina	
	Rhode Island	

by the 10 Compacts to establish new LLRW facilities. Illinois and Kentucky joined together in 1984 to form the Central Midwest Interstate LLRW Compact. The Compact tried to establish a new LLRW facility near Martinsville, Illinois in the 1990s. The site was rejected in 1992 by the Facility Siting Commission because of flaws in the review process, and a loss of public confidence in the site-evaluation process.

The Southwest Compact attempted to LLRW site in Ward Valley, California. In 2002, the California Legislature blocked that effort. US Ecology attempted to site a new facility in Nebraska for the Central Compact, but that effort was halted by the Nebraska Legislature. In 1999, the Governor of Nebraska signed legislation to remove Nebraska from the Central Compact. By 1998, the Appalachian Compact suspended the siting process. The Midwest Compact has also halted their siting process. The Compacts that had access to one of the three remaining LLRW facilities had little motivation to evaluate new sites, and none were proposed.

4.8 The Texas Compact Disposal Facility

In 2011, the Texas Compact Disposal Facility opened (Fig. 4.8). Located in Andrews County, Northwest Texas, this new facility is the first one established since passage of the LLRWPA. The new facility is operated by Waste Control Specialists and is located in an arid climate. The site geology consists of red-colored clay which is the upper part of the Triassic-age Dockum Group. At the site, this clay layer is more than 61.0 m (200 feet) thick with interbedded layers of siltstone and sandstone. The facility has been licensed to accept Class A, B, and C LLRW by the Texas Commission on Environmental Quality. The Texas House of Representatives voted to allow Waste Control Specialists to accept LLRW from 36 other US States that were not part of the Texas Compact. Therefore, because many States that did not have a disposal option for Class B and C wastes after 2008 when Barnwell restricted access to only Compact states, the new Andrews facility solved this waste management issue.

LLRW are placed in 3.0-m (10-foot) tall, 30.5-cm (1-foot) thick concrete canisters and buried 9.1–30.5 m (30–100 feet) below the

Figure 4.8. Texas Compact Disposal Facility. Available at: http://www.wcstexas.com/facilities/compact-waste-facility [Accessed 27 January 2020]. Used with permission of Waste Control Specialists.

surface in cells lined with a geomembrane across the red clay formation. Spaces between the containers are grouted to prevent shifting and to preserve the integrity of the containers. As the cells are filled, they are covered by more than 91.4 m (300 feet) of liner material and clay and, the surface will be restored to its natural state.

From 2012 to 2017, the Texas Compact Waste Facility disposed only 0.36% of the total volume of LLRW disposed in the US (Table 4.7), and about 12.1% of the total amount of radioactivity associated with the wastes disposed during the same period. Given a longer disposal period relative to the older facilities coupled with the Facility's ability to accept Class A, B, and C wastes, the Texas facility is likely to have a major impact on LLRW distribution in the 21st century.

4.9 Greater Than Class C Wastes

GTCC wastes are considered by the US NRC as not acceptable for shallow-land disposal and must be disposed in a geological repository or an alternative proposed by US DOE and approved by the US NRC. There is no disposal site currently available for GTCC LLRW. Many CTCC wastes are currently stored at the site where they were generated. In 2018, US DOE issued an Environmental Impact Statement for the disposal of GTCC LLRW at Waste Control Specialist Federal Waste Facility in Andrews County Texas (US DOE, 2018). The Federal Waste Facility is next to the Texas Compact Disposal Facility.

4.10 US DOE Low-Level Radioactive Wastes

The US DOE also uses shallow-land disposal for the disposition of LLRW. However, US DOE defines LLRW as all radioactive waste that does not fall into other classifications such as high-level, transuranic, or spent nuclear fuel (US DOE, 1999). As a consequence, US DOE does not classify and manage LLRW by class as does the US NRC.

Most US DOE-generated LLRW are managed on the site where they were produced. US DOE has six sites with active disposal for LLRW and mixed wastes generated by past and current activities such as decommissioning and site cleanup (Table 4.9). A major source of LLRW has been from the remediation of sites that were either part of or impacted by the Nuclear Weapons Complex (Chapter 9). LLRW

Table 4.9. US Department of Energy LLRW disposal areas.

Location	Name of current disposal area
Hanford Site	Low-Level Burial Grounds (200 West and East Area)
Savanna River Site	E-Area Low-level Radioactive Waste Disposal Facility
Nevada National Security Site (formerly the Test Site)	Area 5 Radioactive Waste Management Site
Los Alamos National Laboratory	Area G of Technical Area 54, Materials Disposal Area
Idaho National Laboratory	Radioactive Waste Management Complex
Oak Ridge Reservation	Environmental Management Waste Management Facility

were placed in on-site disposal cells that were constructed during site cleanup. It has been concluded that there is sufficient capacity for the disposition of US DOE LLRW to at least 2070 (GAO, 2000).

4.11 Review Questions

1. Name five examples of solid LLRWs.
2. Provide five examples of neutron-activation products and the radionuclide(s) to which they decay.
3. What is DIS?
4. What is a Class A LLRW as defined by the US Nuclear Regulatory Commission?
5. A generator has one 55-gallon container of soil that is contaminated with 3,000 pCi/g ^{238}Pu, 6,000 pCi/g ^{226}Ra, 5,000 pCi/g ^{238}U, 1,100 pCi/g ^{235}U, 5,000 pCi/g ^{234}U, 8,000 pCi/g ^{137}Cs, and 5,000 pCi/g ^{90}Sr. The density of the soil is 1.6 g/cm^3. The disposal facility can accept only Class A waste. Can they accept this container? (borrowed from Energy*Solutions*, 2015).
6. What is the difference between the US Environmental Protection Agency's term "Low-Activity Radioactive Waste" and the US Nuclear Regulatory Commission's term "Low-Level Radioactivity Waste?"
7. What radionuclides are usually in NORM?
8. What were the operational mistakes made at the Maxey Flats Disposal Site and Sheffield LLRW Disposal Facility that led the premature closure of these early disposal facilities?
9. Tritium was detected in shallow groundwater at the Beatty Low Level Radioactive Waste Site, Sheffield LLRW Disposal Facility, and the Barnwell Disposal Facility. Why was tritium so often detected? Why does it come from? (Hint: the reader may need to consult with Chapter 2).
10. The Clive Disposal Facility in Utah has accepted the largest volume of LLRWs of all the current facilities in the US, but not the largest amount of radioactivity. Why?

Bibliography

Adams, J. A. and Rogers, V. L. (1978). *A Classification System for Radioactive Waste Disposal — What Waste Goes Where?* US Nuclear Regulatory Commission (Report number NUREG-0456).

Audi, G., Bersillon, O., Blachot, J., and Wapstra, A. H. (2003). The NUBASE Evaluation of Nuclear and Decay Properties. *Nuclear Physics A*, 729, pp. 3–128.

Barquero, R., Agulla, M. M., and Ruiz, A. (2008). Liquid Discharges from the Use of Radionuclides in Medicine (Diagnosis). *Journal of Environmental Radioactivity*, 99, pp. 1535–1538.

Black, W. D., Solomon, B. J., and Harty, K. M. (1999). *Geology and Geologic Hazardous of Tooele Valley and the West Desert Hazardous Industry Area, Tooele County, Utah.* Special Study 96, Utah Geological Survey.

BLWM. (2007). *Commercial Low-Level Radioactive Waste Management in South Carolina.* Bureau of Land and Waste Management, South Carolina Department of Health and Environmental Control (Report CR-000907).

Contreras, J. (1992). In the Village Square: Risk Misperception and Decision Making in the Regulation of Low-Level Radioactive Waste. *Ecology Law Quarterly*, 19, pp. 481–545.

Energy*Solutions* (2015). *EnergySolutions Clive, Utah Bulk Waste Disposal and Treatment Facilities Waste Acceptance Criteria.* Revision 10.

EPRI. (2014). *Assessment of Radioactive Elements in Coal Combustion Products.* Electric Power Research Institute, report number 3002003779, EPRI, Palo Alto, California.

GAO. (2000). *Low-Level Radioactivity Wastes. Department of Energy has Opportunity to Reduce Disposal Costs.* U.S. General Accounting Office (Report GAO/RCED-00-64).

Gephart, R. E. (2010). A Short History of Waste Management at the Hanford Site. *Physics and Chemistry of the Earth*, 35, pp. 298–306.

ICRP. (1977). *Recommendations of the International Commission on Radiological Protection.* International Commission on Radiological Protection Publication 26.

ICRP. (1979). *Limits for Intakes of Radionuclides by Workers.* International Commission on Radiological Protection Publication 30.

IEMA. (2009). *Site History and Environmental Monitoring Report for Sheffield Low-Level Radioactive Waste Disposal Site.* Illinois Emergency Management Agency, Division of Nuclear Safety.

INEL. (1994). *Directions in Low-Level Radioactive Waste Management: A Brief History of Commercial Low-Level Radioactive Waste Disposal.* Idaho National Engineering Laboratory, U.S. Department of Energy (Report DOE/LLW-103, Rev. 1).

Kelly, W. R. (1987). A Modeling Study of Geochemical Interactions at the Sheffield, Illinois Low-Level Radioactive Waste Disposal Site. *Nuclear and Chemical Waste Management*, 7, pp. 191–199.

Ledoux, M. R. and Cade, M. S. (2002). Licensing and Operations of the Clive, Utah Low-Level Containerized Radioactive Waste Disposal Facility. A Continuation of Excellence. *Waste Management 2002 Conference, February 24–28, 2002 Tucson, Arizona.*

Ryan, M. T., Lee, M. P., and Larson, H. J. (2007). *History and Framework of Commercial Low-Level Radioactive Waste Management in the United States.* Advisory Committee on Nuclear Waste, U.S. Nuclear Regulatory Commission, ACNW White Paper (Report number NUREG-1853).

Saling, J. H. and Fentiman, A. W. (2002). *Radioactive Waste Management.* Taylor & Francis, New York.

US EPA. (2017). *Fourth Five-Year Review Report for Maxey Flats Disposal Site. Fleming County, Kentucky.* US Environmental Protection Agency, Atlanta Georgia.

US EPA. (2019). *Radiation Protection.* US Environmental Protection Agency. Available at: https://www.epa.gov/radiation [Accessed 29 January 2020].

US DOE. (1999). *Radioactive Waste Management Manual.* U.S. Department of Energy (Report DOE M 435.1-1).

US DOE. (2018). *Environmental Assessment for the Disposal of Greater-Than-Class C (GTCC) Low-Level Radioactive Waste and GTCC-Like Waste at Waste Control Specialists, Andrews County, Texas.* US Department of Energy (Report number DOE/EA-2082).

US NRC. (1981). *Draft Environmental Impact Statement of 10 CFR Part 61 Licensing Requirements for Land Disposal of Radioactive Waste. Main Report.* US Nuclear Regulatory Commission (Report number NUREG-0782, Vol. 2).

US NRC. (2019). *Nuclear Materials.* US Nuclear Regulatory Commission. Available at: https://www.nrc.gov/materials/sp-nucmaterials.html [Accessed 29 January 2020].

Wheeler, M. L. and Dragonette, K. (1981). *Low-Level Waste Disposal Methodologies.* Los Alamos Scientific Laboratory (Report LA-UR-81-1385).

Wilshire, H. G. and Friedman, I. (1999). Contaminant Migration at Two Low-Level Radioactive Waste Sites in Arid Western United States — a Review. *Environmental Geology*, 31, pp. 112–123.

Wilson, K. S. and Lyons, B. E. (1991). *Ground-Water Levels and Tritium Concentration at the Maxey Flats Low-Level Radioactive Waste Disposal Site near Morehead, Kentucky, June 1984 to April 1989.* US Geological Survey (Water-Resources Investigations Report 90-4189).

WNA. (2019). *Naturally-Occurring Radioactive Materials (NORM).* World Nuclear Association. Available at: http://www.world-nuclear.org [Accessed 29 January 2020].

Chapter 5

Management of Used Nuclear Fuel

"The sea–the sea," I cried. "Yes," replied my uncle in a tone of pardonable pride; "The Central Sea."

— Henry Lawson in *Journey to the Center of the Earth* by Jules Verne

5.1 Introduction

The need to manage used nuclear fuel has resulted in a number of diverse approaches. During the late 1970s and early 1980s, the National Aeronautics and Space Administration (NASA) in the US studied the feasibility of launching radioactive wastes into orbit around the earth or the moon or sending them directly to the sun (Boeing Aerospace Company, 1982). Because of the risks and consequences of launch failures coupled with the economic costs of sending radioactive wastes into space, the concept was abandoned by NASA. Another approach that was once considered is ice-sheet disposal in which heat-generating wastes are placed on stable ice sheets such as those in Greenland and Antarctica. The hot waste containers would melt the ice then sink downward and become buried. The Antarctic Treaty of 1959, however, bans the disposal of radioactive waste on the continent. Moreover, this approach would require considerable international cooperation. The option of placing used nuclear fuel in a geological repository is discussed in detail in Chapter 6. This chapter focuses on four approaches that have been used in the past or are currently being used: ocean disposal, wet-pool disposal, dry cask storage, and reprocessing used fuel.

5.2 Ocean Disposal

Radioactive wastes were once routinely disposed in the world's oceans. Ocean disposal is essentially a technique that relies on dispersion and dilution rather than long-term containment (Calmet, 1989) as would be expected by placing the wastes into a deep, geological repository. From 1946 to about 1993, thirteen countries used ocean disposal for a variety of radioactive wastes (Table 5.1). Although detailed records were not always made about the type, volume, or activity of the wastes, most of the disposed wastes were low-level wastes (discussed in Chapter 4). Relatively large objects, however, included reactors, some of which contained used nuclear fuel. Ocean disposal occurred at about 80 locations in the Arctic, Atlantic, Pacific Oceans, the Baltic Sea, the Kara Sea, the Barents Sea, and the Sea of Japan (Fig. 5.1). The depth of disposal ranged from 11 to 5,310 m, and it has been estimated that a total of 85,100 TBq (2,300 kCi) of radioactive waste were disposed in oceans from 1946 to 1993 (IAEA, 1999). The United Kingdom and the former USSR accounted for 87% of the total. About 99% of the total radioactivity is from beta and gamma sources (Calmet, 1989).

Table 5.1. Ocean disposal of radioactive waste from 1946 to 1993 (IAEA, 1999).

Country	Activity of the wastes (TBq)	Period of disposal
USSR + Russia	39,246	1959–1993
United Kingdom	35,088	1948–1982
Switzerland	4,419	1969–1982
USA	3,496	1946–1970
Belgium	2,120	1960–1982
France	354	1967–1969
Netherlands	336	1967–1982
Japan	15.1	1955–1969
Sweden	3.2	1959, 1961, 1969
New Zealand	1.0	1954–1976
Germany	0.2	1967
Italy	0.2	1969
South Korea	Not reported	1968–1972

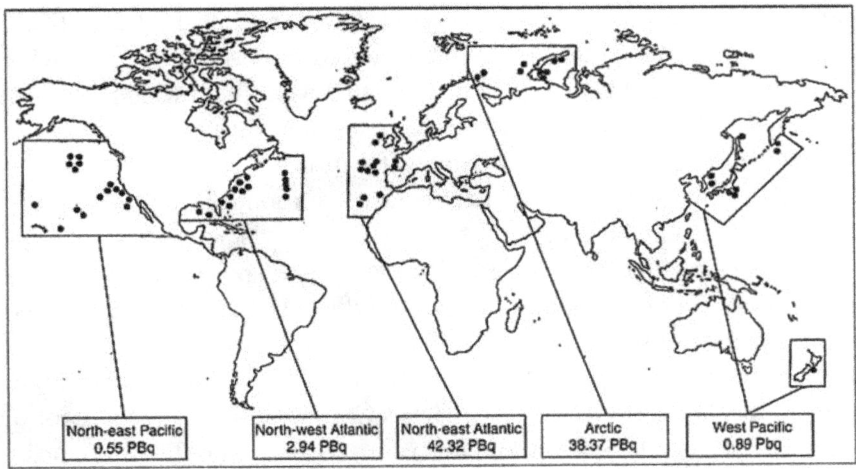

Figure 5.1. Locations of areas used for ocean disposal of radioactive wastes, and an estimate of the total amount of activity disposed in each of the five regions at the time of discharge (IAEA, 1999). Used with permission of the International Atomic Energy Agency.

The Kingdom of Sweden, for example, was among the countries that disposed low-level wastes in the ocean (see Chapter 10). In 1959 and 1961, they disposed of 230 metal drums in the Baltic Sea at a depth of 500 m, containing a total activity of 14.8 GBq (0.4 Ci). In 1969, they disposed of 2,895 metal drums that had been stabilized with concrete and contained a total activity of 3.23×10^3 GBq (86.9 Ci) at a depth of 4,000–4,600 m in the north-east Atlantic (IAEA, 1999).

Bradley (1997) and Yablokov (2001) provided a detailed account of the history of ocean disposal by the former USSR. The official position of the USSR was that they were not using ocean disposal for radioactive wastes. Yablokov concluded that it was not possible to determine accurately the total activity of all the radioactive wastes that were illegally disposed. He further concluded that the available information indicated that a total of 10 reactors with used nuclear fuel from nuclear submarines and the icebreaker Lenin were dumped into the Arctic Ocean. The total activity of objects with used nuclear fuel in the Arctic Ocean has been estimated as 36.88 PBq (996.6 kCi) (IAEA, 1999). Yablokov further added that some of the reactor compartments containing used nuclear fuel were filled with a furfurol (C_4H_3OCH)-based mixture to prevent contact with sea water for up to 500 years. Yablokov added that activation products such as ^{60}Co were also present in some of the reactor components that were sent to the bottom of the sea.

The option for legal international ocean disposal of radioactive wastes no longer exits. Calmet (1989) and the IAEA (1999) described the sequence of events that lead to the London Dumping Convention in 1973, which went into full force in 1975 to stop the disposal of high-level wastes. The passage of the Basel Convention on the Control of Transboundary Movements of Hazardous Wastes and Their Disposal of 1989 and MARPOL 73/78 (the International Convention for the Prevention of Pollution from Ships, 1973 as modified by the Protocol of 1978) added additional restrictions on waste transportation and ocean disposal. The momentum created by a sequence of revisions that lead to a complete ban by 1993. The Russian Federation then ceased ocean disposal. The other 12 countries had ceased ocean disposal by 1982 or earlier (Table 5.1).

A major motivation for phasing out ocean disposal was the concern about adverse impacts on marine life, and the potential to contaminate the marine food chain which of course would include sea food for human consumption. Bioconcentration is a process in which a chemical dissolved in solution becomes concentrated in the tissues of an organism living in the solution. A bioconcentration factor (BCF) is defined as

$$\mathrm{BCF} = \frac{C_{\mathrm{o}}}{C_s} \tag{5.1}$$

where C_{o} is the concentration of the chemical measured in the organism and C_s is the mean concentration of the chemical in solution at equilibrium with that retained by the organism. If the units of C_{o} and C_s are the same, the resulting BCF is unitless.

BCFs are typically measured under laboratory conditions using aquaria-based procedures in which the attainment of equilibrium can

Table 5.2. Bioconcentration factors (unitless) for different groups of marine organisms in Arctic seas (Templeton *et al.*, 1997).

Group	^{241}Am	^{137}Cs	^{210}Po	^{239,240}Pu	^{90}Sr
Phytoplankton	200,000	20	30,000	100,000	3
Zooplankton	2,000	30	30,000	100,000	1
Mollusks	20,000	30	30,000	3,000	1
Crustaceans	500	30	10,000 to 30,000	300	2

be documented by collecting samples of the liquid phase. BCFs are also estimated by collecting aquatic and water samples in the field, and assuming that steady state conditions existed at the time of sampling.

Templeton *et al.* (1997) provided a summary of bioconcentration factors for marine life living in arctic waters. A portion of that compilation is shown in Table 5.2. These factors indicate that strontium-90, and cesium-137 will not be significantly bioconcentrated by phytoplankton, zooplankton, mollusks or crustaceans. There is the potential, however, for ^{241}Am, ^{210}Pu, and ^{239,240}Pu to be enriched in the tissues of each food-web group. The actual concentrations in the tissues would depend on the concentration in the sea water. For example, zooplankton living in water containing a constant concentration of 1 μg/kg of plutonium may contain about 100 mg/kg of plutonium under equilibrium conditions.

Although there is the potential that the marine food chain could be contaminated from the ocean disposal of radioactive wastes, there is a lack of field data that support the assumption that significant contamination occurred before or after the ban on ocean disposal. Baxter *et al.* (1995) summarized the outcomes of field studies that were conducted to assess the impacts of ocean disposal. They reported that near-surface core samples collected in the Kara Sea disposal area contained less than 1.0 Bq/kg of ^{238}Pu and $^{239+240}$Pu. Near-surface core samples collected in the Yenisei Gulf of the Kara Sea contained cesium-137. One core sample contained between 70 and 100 Bq ^{137}Cs/kg, but the source(s) of the cesium was not certain. They also reported that sea-water samples collected *in situ* above the ocean bed at the north-east Atlantic disposal site contained larger concentrations of ^{238}Pu, $^{239+240}$Pu, ^{241}Am, ^{137}Cs, and ^{14}C than samples collected in the control site. The levels of $^{239+240}$Pu, for example, were $\leq 20\,\mu$Bq/L. Baxter *et al.* (1995) concluded that although there have been areas of localized enhancements in radionuclides that implicate disposed wastes as the source, they were radiologically insignificant.

Two other field studies were summarized in IAEA (1999). One was conducted by the US Environmental Protection Agency in the north-east Pacific and north-west Atlantic Ocean disposal areas. Deep-sea-water samples, sediments samples, and deep-sea organisms did not contain levels of radionuclides greater than those attributable to nuclear weapons fallout. A joint Russian–Norwegian study reported that sediment samples collected next to containers in the Kara Sea yielded elevated levels of ^{60}Co, ^{90}Sr, ^{137}Cs, and $^{239+240}$Pu.

Togawa *et al.* (1999) collected surface and bottom water samples in a former disposal area and a background area in the Sea of Japan. The disposal area had received liquid radioactive wastes from the former USSR, and low-level solid wastes from Korea. The researchers found no significant differences between the two areas based on the measured activities of ^{90}Sr, ^{137}Cs, and $^{239+240}$Pu.

Field data suggest that global-scale ocean disposal without any regulatory framework did not result in widespread contamination of sea water or the marine food chain. Additional studies on environmental impacts were, however, recommended by Templeton (1997). Deep ocean disposal would have the advantage of isolating used nuclear fuel from terrorists and rogue nations.

Although ocean disposal in the past was essentially a matter of dumping waste containers and relatively large objects at relatively random locations on the bottom of the sea, sub-seabed sequestration is an untested approach that could be considered. Sub-seabed sequestration is the engineered placement of waste containers in bore holes with clay formations *beneath* the seabed. Hollister *et al.* (1981) may be given credit for first proposing sub-seabed sequestration. They outlined an approach of drilling holes into stable clay formations that cover about 30% of the sea floor. They further concluded that sub-seabed sequestration should be exempt from the London Convention because the wastes would not be "dumped."

McAllister (2013) proposed that sub-seabed sequestration should be evaluated by conducting a pilot-scale project by the US. He proposed that waste containers could be disposed at depth of 4,000 to 5,000 m in the abyssal plain within US Exclusive Economic Zone. McAllister proposed drilling boreholes for the insertion of titanium or stainless-steel containers, followed by a concrete plug. He further generalized that the site selection process would operate with similar criteria as those in siting a geological repository such as avoiding areas of seismic or tectonic activity.

Case study: Subduction-zone disposal

Subduction-zone disposal is a variation of seabed sequestration. The overall approach is to place radioactive wastes in steel containers, then place the containers into excavations in an oceanic seabed or tectonic plate that is actively moving under or being subducted by a continental plate. With time, the radioactive wastes would become subducted by the continental plate. With increasing burial depth, pressures, and temperature, the wastes become dispersed while decaying. The concept of subduction-zone disposal has been patented (Baird, 1991), but it remains untested and subject to debate.

5.3 Wet-Pool Disposal

Wet-pool disposal is the placement of used nuclear fuel assemblies under water. The pools are typically made of reinforced concrete walls with stainless steel liners. The storage pools vary in size but are generally large enough to store the assemblies vertically in water that is about 12 m (40 feet) deep (Werner, 2012). The purpose of the water is to help dissipate the heat given off by the used fuel after it is removed from the reactor, and to provide shielding for power plant workers from occupational exposure.

The use of wet-pool disposal requires that there must always be available a source of makeup water, and electrical power to circulate the water in the pool to avoid it from boiling away and exposing the fuel assemblies. IAEA (2009) generalized that the temperature of the water should be less than 50°C. The pool water is cooled by heat exchangers and passed through ion exchange units to remove corrosion products. Radiolysis of the pool water creates another management concern:

$$2H_2O + \text{ionization radiation} \rightarrow H_2O_2 + H_2 \uparrow \tag{5.2}$$

Because gaseous hydrogen is a potential product, its accumulation increases the risk of explosion. Because of this potential, the air quality of the room containing the pool must be monitored. Some storage pools have leaked tritium into groundwater (Werner, 2012). IAEA (2009) recommended that a leak collection system be an integral component of a storage pool.

Between 2010 and 2030 some 400,000 tonnes of used fuel is expected to be generated worldwide, including 60,000 tonnes in North America and 69,000 tonnes in Europe (WNA, 2018). WNA (2018) estimated that about 90% of this amount is in storage ponds and has been for decades. Of the 67,440 metric tonnes of commercial used fuel in the US and used fuel that is under control of the US Department of Energy, about 73% of it is currently stored in pools (Werner, 2012). In Sweden, all used fuel is currently managed by wet-pool storage. Sweden is unique in that it stores used fuel in pools that are located in an underground repository (see Chapters 6 and 10).

As the need for storage capacity at each power plant increases, some plants re-rack the stored fuel. Re-racking is a process of replacing the racks which are initially used for placement in the pools with higher-density racks that are lined with neutron-absorbing panels to ensure subcriticality. The absorbing medium is made with some type of boron compound. Re-racking can increase storage capacity of a pool by 40–100% (IAEA, 2009). Despite the increased capacity resulting from re-racking, pools have a finite capacity, and the option of building new pools at most power reactors is not practical. Shelton and Bracey (2014) estimated that, on a global scale, about 80% of the total capacity of storage pools is filled with used fuel. At some specific power plants in the US, Taiwan, Switzerland, Spain, South Africa, and Germany, the lack of available capacity forced the operators to seek alternatives. Because of a lack of a geological repository for long-term disposal, an interim solution evolved called dry cask storage.

5.4 Dry Cask Storage

Dry cask storage may be defined as the interim placement of used nuclear fuel into some type of multi-layer cask system. The core of the cask system typically consists of an inner steel canister. The canister contains a steel basket that provides structural support and heat dissipation for the used fuel. Depending on the purpose of the cask, the canister may be placed inside a concrete overpack. The transfer of the fuel into a dry storage system occurs after it has been cooled by wet-pool storage for 5–10 years. Dry storage casks are typically constructed in a cylindrical shape and stored horizontally in above-ground concrete bunkers called a Horizontal Storage Module (HSM) or stored outside vertically on thick concrete pads, or vertically in protective buildings (Werner, 2012). Once a dry cask is filled, sealed, and pressurized with an inert gas, it can provide passive cooling by natural convection of air through the outer concrete shell. The term "cask," however, needs to be expanded (Green *et al.*, 2013).

5.4.1 *Canister*

A sealed inner container employed in many used fuel storage and transport cask systems. These canisters both secure the physical location of used fuel assemblies within a transfer, storage, or transport cask, and provide a mechanism for integrating different types and numbers of fuel assemblies with standardized transfer, storage, or transport casks.

5.4.2 *Transfer Cask*

A used-fuel cask that is employed strictly for transferring used nuclear fuel (individual assemblies or canisters) into storage or transport casks.

5.4.3 *Storage Cask*

A used-fuel cask that is employed strictly for storage of used-nuclear fuel.

5.4.4 *Transport Cask*

A used-fuel cask that is employed strictly for off-site shipping (by road or rail) of used nuclear fuel.

5.4.5 *Dual-Use Cask*

A used-fuel cask that may be used for both storage and transportation of used nuclear fuel. Dual-use or dual-purpose casks are designed and licensed for dry storage and for transportation to a geological repository. Dual-purpose casks may elicit greater public acceptance because the name implies that the cask will be moved eventually, and not become a de facto permanent disposal practice. Shelton and Bracey (2014) cautioned, however, that dry storage casks will likely be in use for decades in many countries. Some type of dry cask storage has been used throughout the world for about 40 years (Table 5.3), and other countries currently planning to use the option in the near future. The first commercial dry storage casks in the US were used in 1986. Werner (2012) reported that 18,102 metric tonnes of used nuclear fuel from utilities and the US Department of Energy are stored in dry casks in the US. Because of the proliferation of dry storage casks, Independent Used Fuel Storage Installations (ISFSI) have been created and are intended to provide interim storage. The design of dry storage cask systems has evolved since its inception. More than 50 different types of dry storage casks have been produced, tested, and approved in the US alone. For example, an early (c. 1987) storage cask was Westinghouse's MC-10, composed of steel, and made

Table 5.3. Summary of the countries that have used or plan to use dry cask storage for used nuclear fuel, and the design capacity of the facilities (from International Atomic Energy Agency — Nuclear Fuel Information Service, 2015).

Country	First used	Design capacity (tonnes of heavy metal)
Argentina	1988	986 and 2,000
Armenia	2000	74 and 80.7
Belgium	1995	1,760 and 2,100
Bulgaria	1984	300 to 600
Canada	1985	Not given
Czech Republic	1995	600 to 1,370
Germany	1995	120 to 3,960
Hungary	1997	850
India	1990	20 and 570
Japan	1995	408 to 6,840 bundles
Republic of Korea	1992	6250
Lithuania	1998	98 bundles
Romania	2003	36,000 bundles/year
Russian Federation	2011	8,130
Slovakia	2017	780
Spain	1993	805
Switzerland	2001	600 and 2,500
Ukraine	1986	2,518
United Kingdom	1979	700
US	1986	26 to 1,112

use of cooling fins attached to the side of the cask. The license for the MC-10 has since expired.

Today, there are five major vendors for dry storage casks: Energy *Solutions*, GNS, Holtec International, NAC International Inc., and AREVA-TN (Table 5.4). The specific design of the casks depends on their intended use. For example, the GNS CASTOR® V/21 (storage in the US) is composed of cast iron with polyethylene rods for a neutron shield.

The fuel basket is composed of cast iron with polyethylene rods for a neutron shield. The fuel basket is composed of borated stainless steel. After sealing, the cask is pressurized with helium

Table 5.4. The major vendors of dry storage casks, and examples of model names currently in use (from Greene *et al.*, 2013).

Vendor	Examples and applications	Cask body composition
Energy*Solutions* (Fuel Solutions™)	W150 (storage cask) W100 (transfer cask) TS 125 (transportation cask) VSC-24 (storage cask)	steel with concrete overpack steel with lead steel with lead vertical concrete cask
GNS Gesellschaft für Nuklear Service	CASTOR and CONSTOR (storage and transportation casks)	iron
Holtec International	HI-TRAC (transfer cask) HI-STORM and HI-STAR (storage and transportation overpacks)	steel with lead steel with lead or concrete
NAC International Inc.	MAGNASTOR and NAC series (storage and transportation casks) MAGNATRAN (transportation casks)	steel with concrete and steel with lead
AREVA-TN (formerly Transnuclear and AREVA is now Framatome)	NUHOMS series (storage and transportation canisters) TN series (storage and transportation casks)	steel

(Green *et al.*, 2013). The fuel basket within the HI-STORM 100 Multi-Purpose Canister (MPC) is composed of stainless steel and Metamic™ panels along the length of the basket. Metamic™ is a neutron-shielding material that was invented by Holtec International, and it is composed of aluminum boron carbide. The panels are welded to the MPC The MPC is then placed into a concrete overpack, sealed, and pressurized with helium (Fig. 5.2). For a NUHOMS® system, the concrete overpack is a stationary, horizontal concrete bunker called a HSM (Fig. 5.3). The transfer cask is placed into the concrete block horizontally, and then the HSM is sealed.

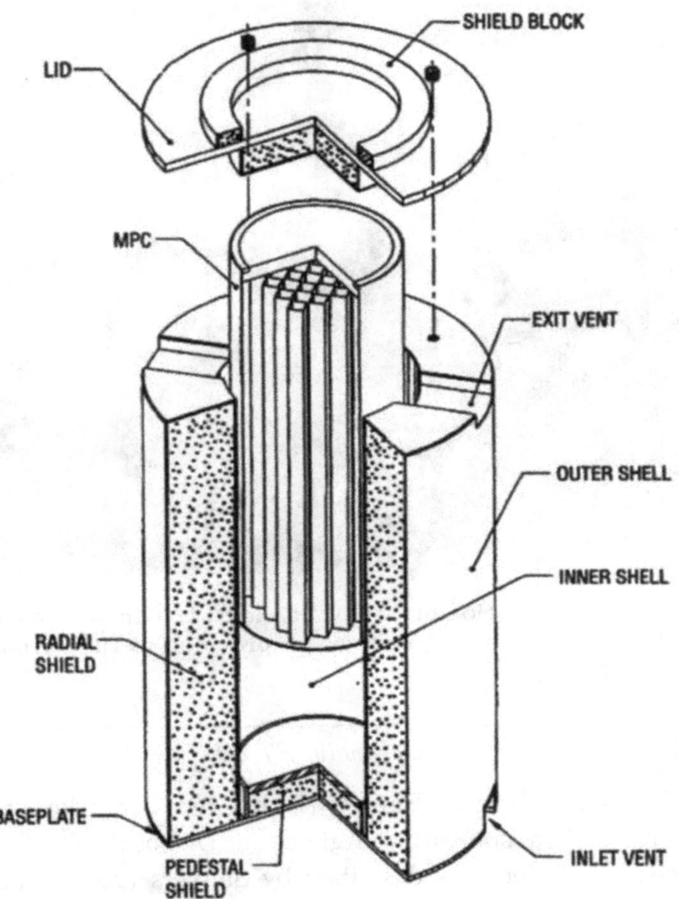

Figure 5.2. **The HI-STORM 100 above-ground overpack and multi-purpose Canister (US DOE 2013).**

Figure 5.3. Construction of a HSM at the R. E. Ginna Nuclear Power Plant, Ontario, New York (photograph provided to the author by the Power Plant).

Case study: STAD

The US Department of Energy (DOE) has been developing a Standardized Transportation, Aging, and Disposal (STAD) waste management system. As described by Sanders (2013), the basic concept of the STAD system is to place used nuclear fuel into a steel canister which is sealed with no need to reopen the canister for future repackaging or transportation. Depending on the need, the canister can be placed in a concrete overpack for interim storage or aging. A second option is to place the canister into a transportation overpack for moving the used fuel to a centralized above- ground storage facility or a geological repository. The STAD concept, however, evolved for use at the Yucca Mountain Nuclear Waste Repository. Because the future of the Yucca Mountain project is uncertain, additional research and development of the STAD system is also in limbo.

5.5 Reprocessing and Advanced Separation Technology

5.5.1 *The Bismuth Phosphate Process*

Used nuclear fuel has always been reprocessed by one method or another somewhere in the world. The first used fuel was created as a feedstock for plutonium in the US. The plutonium was extracted to make nuclear weapons to help end the war against Japan. The method selected to extract the plutonium was called the bismuth phosphate process. It was first used on a large scale at the Hanford Site in Washington State which was established in 1943 as part of the Manhattan Project.

The used fuel was first cooled for about 40 days. Then the aluminum cladding was dissolved in a hot sodium hydroxide–sodium nitrate solution (Gray, 1999). The fuel was then dissolved in a 60% nitric acid solution. The plutonium that was dissolved was reduced to Pu^{4+} by the addition of sodium nitrite. The nitrite oxidized to nitrate, and the resulting electrons reduced the plutonium. Then bismuth nitrate and phosphoric acid were added to co-precipitate the plutonium:

$$\begin{aligned} &\mathrm{Pu} \xrightarrow{\mathrm{NaNO2}} \mathrm{Pu^{4+}} + \mathrm{Bi(NO_3)_3} + \mathrm{H_3PO_4} \rightarrow \\ &\mathrm{BiPO_4(Pu^{4+})} \downarrow + 3\mathrm{NO_3^-} + 3\mathrm{H^+} \end{aligned} \tag{5.3}$$

The precipitate was collected by centrifugation, then dissolved in nitric acid. The final product was refined to remove impurities such as fission products by numerous dissolution–precipitation steps. Plutonium yields improved with time, reaching 90% (Gray, 1999). A major limitation of the bismuth phosphate process, however, was that it did not recover uranium (Choppin, 2002). Moreover, it appeared that plutonium could only be recovered in batches, and that the process created relatively large volumes of waste solutions.

The bismuth phosphate process at Hanford was replaced by a number of liquid–liquid solvent-based extractions. The solvent chosen has a greater chemical affinity for the solutes (such as uranium and plutonium) than the acidic, aqueous phase created by the dissolution of the used fuel in nitric acid. In this context, three terms need to be defined (derived from Simpson and Law, 2010):

- **Extraction:** The transfer of solute(s) from one liquid phase to another. Typically, the transfer of solutes from an aqueous, hydrophilic phase to a hydrophobic phase composed of an organic solvent.
- **Scrubbing:** The purification of an intermediate product such as the removal of fission products from the solvent that were co-extracted with uranium and plutonium.
- **Stripping:** The back extraction; the transfer of solute(s) from the hydrophobic phase back into a hydrophilic aqueous phase for additional separations and recovery.

The first continuous solvent-based extraction process that replaced the bismuth phosphate process at Hanford was the REDOX process. This extraction process used methyl isobutyl ketone (MIK, $[(CH_3)_2CHCH_2COCH_3]$) as the hydrophobic solvent. MIK was mixed with the acidic, aqueous phase (Simpson and Law, 2010). Uranium (VI) and plutonium (IV and VI) ions react with MIK:

$$Pu^{4+} + 4NO_3^- + 2MIK \rightleftharpoons Pu(NO_3)_4(MIK)_2 \qquad (5.4)$$

making them soluble in the organic hydrophobic phase (Castano, 2011). Fission products and plutonium (III) are relatively insoluble in MIK. The major disadvantage of the REDOX process was that MIK would decompose in nitric acid (Choppin, 2002).

5.5.2 *PUREX*

The Plutonium URanium Extraction (PUREX) process was invented in 1947 in the US and is currently the global de facto standard for used nuclear fuel reprocessing. Since about 1947, eleven countries have used, or are currently using the PUREX process in the production of nuclear weapons or mixed-oxide fuel for power generation (Table 5.5).

Table 5.5. Summation of the countries that have processed used nuclear fuel.

Country	Process	Operational period
Belgium	PUREX	1966–1974
China	PUREX	1968–1973, 2004
Germany	PUREX	1971–1990
France	PUREX, DIAMEX,[a] SANEX[1]	1976–present
Italy	EREX (URanium Extraction)	1970–1987
India	PUREX	1964–present
Japan	PUREX	1972–present
Pakistan	PUREX	1982–present
Russia	PUREX	1940s–present
United Kingdom	PUREX	1964–present
USA	Bismuth phosphate, REDOX, PUREX	1944–2002

Note: This compilation combines all forms of used fuel (weapons production, research, and power production), and all processing sites during the operational period. The data were compiled from the World Nuclear Association. Available at: http://www.world-nuclear.org/ [Accessed 29 January 2020].
[a]Raffinate treatment procedures. DIAMide EXtration and Selective Actinide EXtraction (see Castano, 2011).

PUREX is a liquid–liquid extraction that separates actinides from used fuel. The acidic, aqueous phase is mixed vigorously with an organic phase composed of 30% tributyl phosphate (TBP) ($(CH_3CH_2CH_2CH_2O)_3PO$) in kerosene or dodecane ($CH_3(CH_2)_{10}CH_3$). The key to understanding the PUREX process is that uranium and plutonium react with TBP and nitrate. This reaction converts the hydrophilic ions into hydrophobic forms:

$$UO_2^{2+} + 2NO_3^- + 2TBP \rightleftharpoons UO_2(NO_3)_2.2TBP \quad (5.5)$$

$$Pu^{4+} + 4NO_3^- + 2TBP \rightleftharpoons Pu(NO_3)_4.2TBP \quad (5.6)$$

The hydrophobic forms of uranium and plutonium then partition from the aqueous phase into the organic hydrophobic phase. In general, metals in the 4+ and 6+ oxidation states are extracted by the PUREX process (Simpson and Law, 2010). Plutonium (III) and other actinide and lanthanides in the 3+ or less oxidation state remain in the aqueous phase. Fission products such as Cs-137 and Sr-90 largely remain in the acidic aqueous phase but can be removed by scrubbing in case a small fraction of them are co-extracted with uranium and plutonium.

5.5.3 *Advanced Versions of PUREX*

The primary objective of developing advanced PUREX processes is to avoid the separation of free plutonium that could pose proliferation risks. Other goals include the separation and recovery of neptunium and technetium (Simpson and Law, 2010). In France, the COEX™ process is being developed. During the "co-conversion process," a uranium–plutonium oxide solid solution[1] ((U, Pu)O_2) is precipitated which is suitable for mixed oxide (MOX) fuel fabrication (Drain *et al.*, 2008). Because plutonium occurs in a solid solution, it poses few proliferation concerns. Drain et al. also indicated that a future goal of COEX™ is to produce neptunium–uranium–plutonium mixed oxide.

UREX is another modified PUREX process designed to prevent plutonium from being extracted independently. The goals for the UREX process are to recover >99.9% of the uranium and >95% of the technetium in separate product streams while rejecting >99.9% of the TRU radionuclides to the raffinate. To meet this requirement, the process was designed to use acetohydroxamic acid ($C_2H_5NO_2$) in the scrub stream which complexes Pu (IV) and Np (IV), preventing them from being extracted. The UREX process has been incorporated in several hybrid processes such as UREX + 1a which is a sequence of four solvent extractions and one ion exchange operation to separate fission products and actinides from the raffinate (see Pereira *et al.*, 2007).

It is beyond the scope of this chapter to review the many aqueous separation methods that are under development for extracting desirable fissionable radionuclides while treating the raffinate waste

[1] A crystalline material in which two or more elements or compounds share a common lattice.

streams, which could potentially lead to improved methods for waste disposal. Major research efforts on advanced reprocessing are ongoing in France, Japan, the United Kingdom, China, Russia, and the US (Simson and Law, 2010). The reader is directed to Castano (2011) for a recent review. None of these advanced aqueous separation methods have been applied to full-scale reprocessing to date (Simpson and Law, 2010).

Case study: TALSPEAK

TALSPEAK (Trivalent Actinide Lanthanide Separation by Phosphorus Extractions and Aqueous Komplexes) is a solvent-extraction process that was designed in the 1960s at Oak Ridge National Laboratory in the US. The purpose of TALSPEAK is to separate actinides lanthanides in the PUREX raffinate. The lanthanides are selectively extracted with di(2-ethylhexyl)phosphoric acid which is a liquid cation exchanger and a chelation agent (Nilsson and Nash, 2008). The actinides are complexed by the addition of polyaminoployacetic acid and remain in the aqueous phase.

5.5.4 *Pyroprocessing*

Pyrochemical separation or pyroprocessing is an alternative to aqueous extractions. It is a high-temperature, selective electrochemical dissolution process that is at the experimental stage. Bascially the electrorefiner contains a mixture of molten lithium chloride and potassium chloride in a temperature range of 500–800°C as the electrolyte. The used fuel is chopped into segments and placed in a basket which is used as the anode (Fig. 5.4). As a current is passed between the anode and the cathode, uranium is oxidized to U(III) at the anode, then reduced to the metallic form at the cathode. Two different types of cathodes are used. One is a solid iron or steel electrode that is used to recover uranium. The other is a liquid cadmium electrode that is used to recover transuranic elements. Electrochemically active fission products are oxidized and enter the electrolyte where they accumulate in a salt mixture (Castano, 2011).

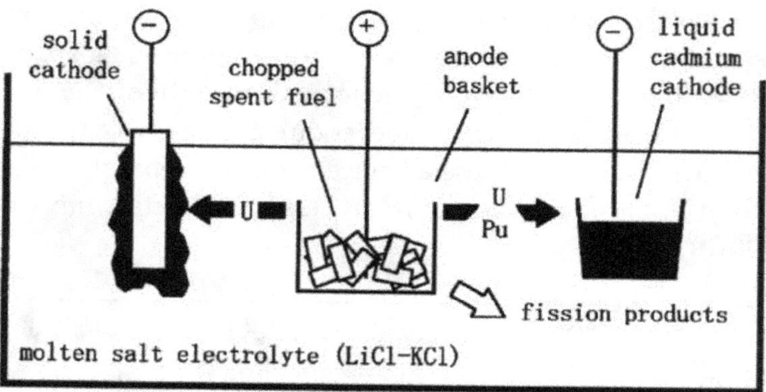

Figure 5.4. Schematic of an electrorefiner with used nuclear fuel (Iizuka *et al.*, 2001). Used with permission of the Organization for Economic Co-operation and Development.

The uranium that accumulates at the solid cathode can then be refined for reuse. There are various electrorefiners and procedures that have been designed and tested at Argonne National Laboratory (US) and in South Korea.

5.6 Review Questions

1. What are the management requirements of operating wet-pool storage?
2. Explain how the PUREX process separates actinides from used nuclear fuel.
3. How do the advanced versions of PUREX address the concerns about plutonium-driven proliferation?
4. Evaluate the feasibility of sub-seabed sequestration of used nuclear fuel.
5. Discuss the various forms of dry storage cask used in managing used nuclear fuel.
6. Why would a dry storage cask be filled with an inert gas prior to final closure?
7. Compared the received risks of wet storage when compared to dry storagecasks (the reader will need conduct additional research).

8. What is the difference between extraction, scrubbing, and stripping during the reprocessing of used nuclear fuel?

Bibliography

Baird, J. R. (1991). *Subductive Waste Disposal.* US Patent 5 022 788. Date issued 11 June 1991.

Baxter, M. S., Ballestra, S., Gastaud, J., Hamilton, T. F., Harms, I., Huynh-Ngoc, L., Kwong, L. L. W., Osvath, I., Parsi, P., Petterson, H., Povinec P. P., and Sanchea, A. (1995) Marine Radioactivity Studies in the Vicinity of Sites with Potential Radionuclide Releases. In *Proceedings of a Symposium (Environmental Impact of Radioactive Releases)*, Vienna, May 8–12, 1995. International Atomic Energy Agency, p. 125–141.

Boeing Aerospace Company. (1982). *Analysis of Space Systems for the Space Disposal of Nuclear Waste Follow-Up Study*, Volume 2 (Technical Report D 180-22777-2).

Bradley, D. J. (1997). Naval Waste Management and Contamination of Oceans and Seas In Payson, D. R. (ed.), *Behind the Nuclear Curtain: Radioactive Waste Management in the Former Soviet Union.* Battelle Press, Columbus, Ohio, pp. 292–324.

Calmet, D. (1989). *Ocean Disposal of Radioactive Waste: Status Report.* International Atomic Energy Agency (Bulletin 4/1989).

Castano, C. H. (2011). Nuclear Fuel Reprocessing. In Krivit, S. B., Lehr, J. H. and Kingery, T. B. (eds), *Nuclear Energy Encyclopedia*, John Wiley & Sons, Hoboken, New Jersey, pp. 121–126.

Choppin, G. R. (2002). Overview of Chemical Separation Method. In Choppin, G. R., Khankhasayev, M. K., and Plendl H. S. (eds.), *Chemical Separations in Nuclear Waste Management*, Battelle Press, Columbus, Ohio, pp. 3–16.

Drain, F., Emin, J. L., Vinoche, R., and Baron, P. (2008). COEX^{TM} Process: Cross-Breeding between Innovation and Industrial Experience. *Waste Management 2014 Conference*, Phoenix, Arizona, USA, February 24–28, 2008.

Gray, L. W. (1999). *From Separations to Reconstitution — A short History of Plutonium in the US and Russia.* Lawrence Livermore National Laboratory (Report UCRL-JC-133802).

Hollister, C. D., Anderson, D. R., and Heath, G. R. (1981). Subseabed Disposal of Nuclear Wastes. *Science*, 213, pp. 1321–1326.

IAEA. (1999). *Inventory of Radioactive Waste Disposals at Sea.* International Atomic Energy Agency (Report IAEA-TECDOC-1105).

IAEA. (2009). *Costing of Spent Nuclear Fuel Storage.* International Atomic Energy Agency (Report Series NF-T-3.5), Vienna, Austria.

Iizuka, M., Uozumi, K., Inoue, T., Iwai, T., Shirai, O., and Arai, Y. (2001). Development of Plutonium Recovery Process by Molten Salt Electrorefining with Liquid Cadmium Cathode. In *Organization for Economic Co-operations and Development-Nuclear Energy Agency 6th International Information*

Exchange Meeting on Actinide and Fission Product Partitioning and Transmutation, Madrid, Spain, December 11–13, 2000.

McAllister, K. R. (2013). Sub-Seabed Repository for Nuclear Waste — A Strategic Alternative — 13102. *Waste Management 2013 Conference*, Phoenix, Arizona, US, February 24–28, 2013.

Nilsson, M. and Nash, K. L. (2008). TALSPEAK Chemistry in Advantaged Nuclear Fuel Cycles. In *The 3rd International ATALANTE Conference, Nuclear Fuel Cycles for a Sustainable Future*, Montpellier, France, May 19–23, 2008.

Pereira, C., Vandergrift, G. F., Regalbuto, M. C., Bakel, A., Bowers, D., Gelis, A. V., Hebden, A. S., Maggos, L. E., Stepinski, D., Tsai, Y., and Laidler, J. J. (2007). Lab-Scale Demonstration of the UREX+1a Process Using Spent Fuel. *Waste Management 2007 Conference*, Phoenix, Arizona, US, February 25–March 1, 2007.

Sanders, C. E. (2013). Review of the Development if the Transportation, Aging, and Disposal (TAD) Waste Disposal System for the Proposed Yucca Mountain Geological Repository. *Progress in Nuclear Energy*, 62, pp. 8–15.

Shelton, C. and Bracey, W. (2014). Perspectives on Interim Storage Solutions of Used Nuclear Fuel in the Long Term. In *Waste Management 2014 Conference*, Phoenix, Arizona, USA, March 2–6, 2014.

Simpson, M. F. and Law, J. D. (2010). *Nuclear Fuel Reprocessing*. Idaho National Laboratory (Report INL/EXT-10-17753).

Templeton, W. L. (1997). Research Needed Relative to Radiological Assessment of the Dumping of Radioactive Wastes in the Oceans. *Marine Pollution Bulletin*, 35, pp. 374–380.

Templeton, W., Harrison, F., Knezovich, J., Fisher, N., and Layton, D. (1997). Bioconcentration of Radionuclides in Marine Food-Web Organisms. In Layton, D., Edson, L.C.D.R.R., Varela, M. and Napier, B. (eds.), *Radionuclides in the Arctic seas from the Former Soviet Union: Potential health and ecological risks, Arctic Nuclear Waste Assessment Program (ANWAP)*, Office of Naval Research (ONR), pp. 4-1–4-12.

Togawa, O., Povinec, P. P., and Petterson, H. B. L. (1999). Collective Dose Estimates by Marine Food Pathway from Liquid Radioactive Wastes Dumped into the Sea of Japan. *The Science of the Total Environment*, 237–238, pp. 241–248.

US DOE (2013). Preliminary Thermal Modeling of HI-STORM 100S-218 Version B Storage Modules at Hope Creek Nuclear Power Station ISFSI, Report number FCRD-UFD-2013-000297PNNL-22552 Rev.1.

Werner, J. D. (2012). *U.S. Spent Nuclear Fuel Storage*. CRS Report for Congress, Congressional Research Service (Report 7-5700).

WNA. (2019). *Processing of Used Nuclear Fuel*. World Nuclear Association. Available at: https://www.world-nuclear.org/information-library.aspx [Accessed 23 November 2019].

Yablokov, A. V. (2001). Radioactive Waste Disposal in Seas Adjacent to the Territory of The Russian Federation. *Marine Pollution Bulletin*, 43, pp. 8–18.

Chapter 6

Geological Repositories

"Do you know, my dear boy, that to reach the interior of the earth we have only five thousand miles to travel?"

— Prof. Von Hardwigg in *Journey to the Center of the Earth* by Jules Verne

6.1 Introduction

A geological repository is an excavation within solid rock for the purpose of long-term containment and isolation of radioactive waste. There is a strong consensus among all major countries that harness nuclear energy that a deep, geological repository is the best option available for managing radioactive wastes (Allan and Nuttall, 1997; Chapman and Hooper, 2012). The purpose of a geological repository is to provide passive protection from the release of radioactive material and to prevent unacceptable health risks to future generations (Chapman and Hooper, 2012).

During the last 40 years, the international concept of geological disposal has been that of a multi-barrier system in which engineered and geological barriers act in concert to isolate the waste by preventing radionuclides from leaching from the wastes and being transported into the biosphere (Chapman and Hooper, 2012). Although the geology and engineering designs vary from one country to the next (see also Chapter 10), these multi-barrier systems can be generalized as:

1. The solid waste itself can be a barrier to leaching — such as solid vitrified wastes.

2. The metal container around the waste form.
3. The overpack around the container.
4. The backfill/buffer surrounding the overpack.
5. The rocks that surround the buried overpack.

The geological environment that is most conducive to meeting the goals of long-term containment and isolation must consist of geological material that is relatively impermeable to groundwater movement that could transport radionuclides. It is also crucial to avoid locations with a potential for significant seismic or near-surface volcanic activity. Other long-term considerations include tectonic processes that could result in the slow deformation of the rocks around the waste packages. Climate change resulting in changes in sea level and future continental glaciation also complicate the site-selection process. Hence, to achieve the goal of long-term containment and isolation, the geology, geochemistry, and hydrogeology of a potential site must be studied in *great* detail. Brookins (1984), Chapman and McKinley (1987), and Roxburgh (1987) are examples of repository citing and design from a geological perspective.

Internationally, geological repositories that currently contain wastes or are planned in the future for low- and intermediate-level radioactive wastes, transuranic wastes, spent nuclear fuel, and high-level reprocessing wastes vary in terms of capacity, depth, design, and type of host rock (see also Chapter 10). The type of host rock available is influenced by the long-term geological history of the area chosen for the site-selection process. The host rock for most planned geological repositories is either igneous and metamorphic rocks — such as granite and gneiss, and sedimentary rocks such as argillites (claystone), evaporates (halite), and consolidated volcanic ash (Chapman and Hooper, 2012).

The design and operation of the repository depends on the waste management philosophy, and the requirements imposed during the site-selection process. The *disposal* of radioactive wastes is the permanent emplacement of the wastes with no intention of moving or retrieving the wastes in the future. *Storage-disposal* is the emplacement of radioactive wastes in a repository with the option of being able to retrieve the wastes later before the final closure of the repository. Being able to remove all or part of the waste packages could be implemented if major problems with the repository were identified later. If spent nuclear fuel was stored in a repository, future generations would have the ability to retrieve the waste packages if reprocessing of used fuel became an economical alternative. Lastly, the spent fuel could be removed if a better alternative becomes available in the future, and the repository could be emptied then decommissioned. It seems unlikely that low- and intermediate-level, transuranic, and high-level radioactive wastes from reprocessing spent fuel would have any value as a resource in the future and would not require retrievability.

There are some potential problems with retrievability. If the repository is not sealed completely to allow for future accessibility, there could be potential groundwater contamination, and the movement of radioactive gases into the biosphere. Having the option of retrievability would also add to the costs of construction and operation of the repository. Regardless of the debate, most countries have required reversibility as part of the repository design. In the US for example, the Nuclear Waste Policy Act of 1982 states that *any* geological repository will be designed and constructed to permit retrieval. The US NRC currently requires that the waste must be retrievable at any time to 50 years after the start of emplacement (§63.111 Performance objectives for the geologic repository operations area through permanent closure in Part 63 — Disposal of High-Level Radioactive Wastes in a Geologic Repository at Yucca Mountain, Nevada (see https://www.nrc.gov/reading-rm/doc-collections/cfr/part063/full-text.html [Accessed 31 January 2020].

6.2 Natural Analogues of a Geological Repository

There are natural analogues in nature that provide evidence that geological repositories can isolate radioactive wastes for long periods of time. Possibly the most well-known is the open-pit uranium mine at Oklo in the Gabonese Republic in Africa. In 1972, routine uranium samples were assayed and found to be depleted with respect to uranium-235. Further study revealed that uranium samples from the depleted vein contained radionuclides that were typically products of nuclear fission but were almost absent elsewhere in the ore body (Cowan, 1976). Eventually, Francis Perrin and other French scientists realized that that the ore body had once contained enough uranium-235 to sustain a chain reaction which consumed a portion of its fuel, then ceased during the Pre-Cambrian Era, about 1.8 billion years ago. Sixteen reactor zones have been identified at Oklo (Fig. 6.1).

The Oklo reactor zone-uranium oxide is an analog for spent fuel. It is surrounded by ferromagnesium clay minerals (illite and chlorite) that may be analogous to an engineered barrier made of clay materials (Brookins, 1990). It has been estimated that the Oklo reactors were operative for about 800,000 years. The long-term presence of some of the fission products within and near to reactor zones supports the potential efficacy of a geological repository with engineered barriers. The Oklo case study, however, must be interpreted cautiously. Many of the fissiogenic radionuclides that were formed about 1.8 billion years ago are absent now because of

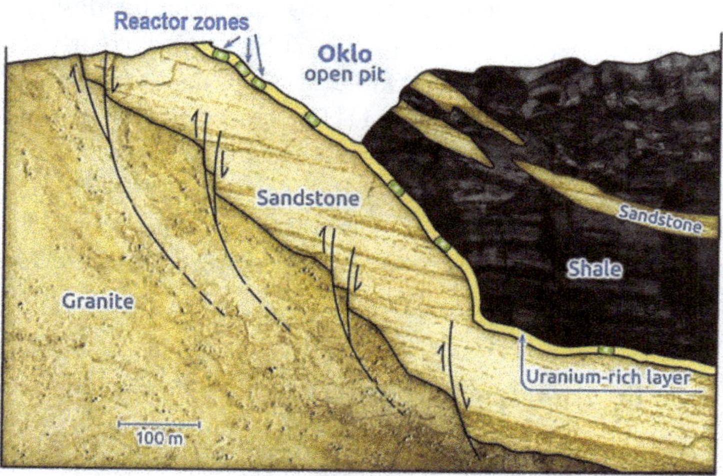

Figure 6.1. A simplified geological cross section of the Oklo uranium deposits (modified from Mossman *et al.*, 2008).

radioactive decay and were assumed to have been present in the past because of the contemporary presence of daughter-decay products. For example, any fissiogenic strontium-90 would have decayed to yttrium-90 which decays to stable zirconium-90 which has been observed in the sandstone near the reactor zone (Table 6.1). The mechanisms assumed to have been responsible for the apparent retention (or absence) of the radionuclides were based on several factors. These factors include geochemical processes such as chemical precipitation–dissolution, sorption–desorption, and diffusion — coupled with estimates and assumptions about groundwater movement and the role of the water to act as a moderator to slow fast neutrons, making them more effective in a fission-chain reaction. Lastly, estimates of the chemical composition of the atmosphere at that time, reactor temperatures, and the role of the mineralogical properties of both the uranium ore and the surrounding sandstone were factored into the analysis. In other words, Oklo was not a controlled experiment that can be repeated.

Despite these complications, some observations have been made (Table 6.1). The uranium deposits were stable for almost two billion years. There was little evidence that the uranium was dissolved and re-precipitated either during the nuclear reactions, or during the two billion years that followed (Berzero and D'Alessandro, 1990). Many of the fission products and actinides remained in place or close to where they formed (Brookins, 1990). The actinides that are compatible with uranium oxide in terms of ionic radii and can form solid solutions with UO_2. Of the fission products, the ones that tended to be retained were less volatile and had ionic radii that were similar to that of uranium. The efficacy of the gangue (commercially worthless material near the ore) as a barrier is more difficult to evaluate. The retained fission products may have eventually been remobilized by ion exchange or dissolution by circulating groundwater during the two billion-year interval. Regardless, the observations at Oklo show that very little of the radionuclide inventory migrated during the 800,000 years of reactor operation, and that most of the radionuclides and/or their decay products have not migrated or have migrated only a few meters (Brookins, 1990).

Table 6.1. Migration of fission products and actinides at the Oklo natural reactor sites.

Radionuclide	Observation
Americium	The yields from the reactions may have been limited.
Barium	Behavior of fissiogenic barite masked by naturally occurring barite.
Bismuth	^{209}Bi was derived from the decay of ^{237}Np. Remained with the host UO_2 possibly as a sorbate (half-life of ^{237}Np is 2.12×10^6 years).
Bromide	Assumed to have migrated from the host UO_2.
Cadmium	Migrated from the host UO_2 possibly by volatilization.
Indium	^{115}In was likely retained (half-life of 4.4×10^{14} years).
Iodine	Fissiogenic iodine was not retained.
Molybdenum	About 80% of the fissiogenic Mo isotopes had migrated from the host UO_2 but were retained as a sulfide mineral within 1.5 m of the reactor zone in the sandstone gangue (the half-life of ^{100}Mo is 8.5×10^{18} years).
Neodymiun	Retained in the host UO_2. Stable, fissiogenic ^{143}Nd present in greater amounts than background levels.
Neptunium	Assumed to have been retained by sorption and solid solution with UO_2 as inferred by the presence of ^{209}Bi.
Niobium	Retained in the host UO_2.
Palladium	Retained in the host UO_2 (the half-life of ^{107}Pd is 6.5×10^6 years).
Plutonium	Assumed to have been retained by solid solution until decay (the half-life of ^{239}Pu is 24,110 years).
Rhodium	Presence of ^{103}Rh greater than background implies that it was retained in the host UO_2 (^{103}Rh is a stable isotope that forms from the fission of ^{235}U).
Ruthenium	Most (70–90%) ^{99}Ru retained in the host UO_2 (likely as a decay product of ^{99}Tc which has a half-life of 2.1×10^5 years).
Strontium	Fissiogeninic strontium may have migrated from the host UO_2. Possibly sorbed by sandstone gangue as inferred by the presence of stable ^{90}Zr.
Technetium	About 60–85% of the ^{99}Tc was retained in the host UO_2.
Tellurium	Most of the ^{126}Te was retained in the host UO_2 (^{126}Te is stable).
Thorium	Retained in the host UO_2.
Xenon	Fissiogenic xenon migrated from the host UO_2.
Yttrium	Likely that most fissiogenic Y was retained by the host UO_2.

Source: From Berzero and D'Alessandro (1990), Brookins (1990), Cowan (1976), and Curtis *et al.* (1989).

There are numerous other geological deposits that provide evidence for the long-term stability of uranium deposits that can be used as a natural analog to geological repositories. Smellie (2009) presented an analysis of 21 uranium ore bodies that have been studied as natural analogs in the context of long-term stability, or as field-scale studies that yielded information on the solubility, movement in groundwater and retardation of radionuclides by geological media.

One such study is the Cigar Lake deposit in Saskatchewan (Fig. 6.2), Canada, the world's richest deposit of uranium. Located at a depth of about 430 m below land surface, it contains from 0.3% to 55% uranium, and is estimated to be about 1.3 billion years old (Cramer, 1995). Its longevity has been attributed to two factors: the chemically reduced nature of the groundwater such that the solubility of uranium is minimized, and the fact that the uranium deposit is encased with a 10- to 50-m thick layer of clay — predominantly illite — that has protected the uranium ore from groundwater flow.

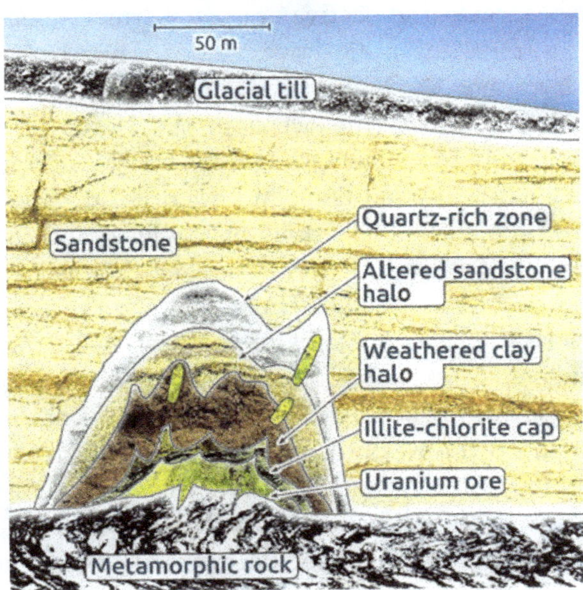

Figure 6.2. Geological cross section of the Cigar Lake uranium deposit in Saskatchewan, Canada (modified from Cameco, 2012).

Cramer (1995) pointed out that the illite layer is a natural analog to layers of bentonite clay which is a barrier in the design of the deep geological repository in Canada.

The Sanerliu uranium deposit in People's Republic of China was offered by Min *et al.* (1998) as a natural analog to what is envisioned for a geological repository of high-level waste in China (see Chapter 10). Estimated to be about 215 million years old (215 Ma), the deposit is at a depth of 160 to 400 m and is hosted by granite. Min *et al.* (1998) proposed that a clay-rich alteration halo that surrounds the ore body is similar to backfill-clay material in a geological repository.

6.3 Repositories Used for Low-Level, Intermediate-Level, and Transuranic Radioactive Wastes

Geological repositories are being used to store low-level, intermediate-level and transuranic radioactive wastes. Former mines are being used as repositories for radioactive wastes. Four examples are in the Federal Republic of Germany. Evaporitic deposits such as salt (halite) precipitated from the ocean in the North German Basin during the Mesozoic Era. The Asse II Repository is a former salt and potash mine located in a salt dome in Lower Saxony (Fig. 6.3). From 1967 to 1978, about 125,800 drums of low-level and intermediate-level waste were disposed in 13 tunnels at depths of about 511, 725, and 750 m (Rempe, 2007). Unfortunately, the Asse mine was excavated close to the boundaries of the host salt dome. Flooding and the slow loss of mechanical stability doomed the Asse site as an acceptable geological repository. In 2010, the German government decided to remove the waste drums during the next 15 years. The decommissioning of the Asse II Repository is currently under study.

The Morsleben Repository is also a former potash and salt mine and is in Saxony-Anhalt in Upper Permian-age salt. Commissioned in 1971, the salt mining operations created an underground volume of space of about 8 million m^3 (Riplens and Biurrun, 2002). From 1971 to 1998, about 36,800 m^3 of low-level and intermediate-level waste

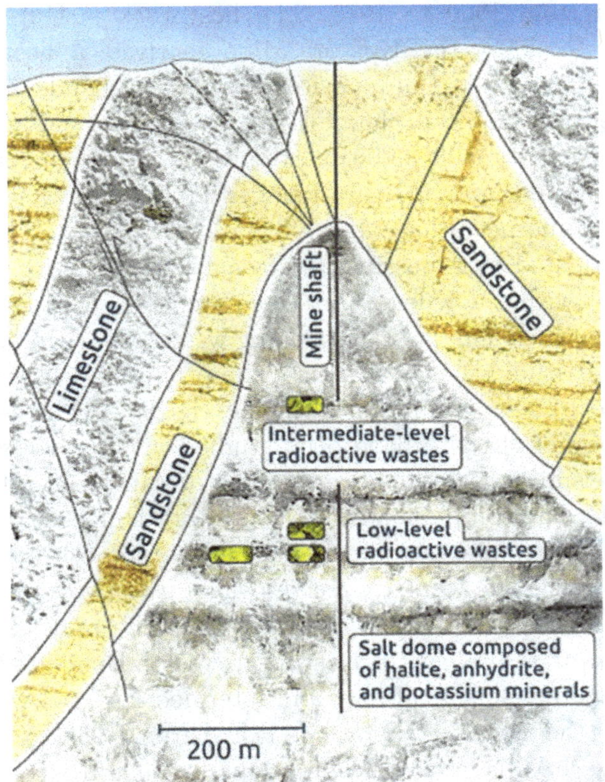

Figure 6.3. Geological cross section of the Asse II Repository in Lower Saxony, Germany (modified from Schwartz, 2009).

were disposed at a depth of about 500 m, including 6,621 sealed-radiation sources (Rempe, 2007). The total activity of the waste was estimated to be 1.2×10^{12} Bq (32 Ci) (Preuss *et al.*, 2002). Much of waste disposed consisted of radionuclides with a half-life that is less than 30 years (Ripkens and Biurrun, 2002). Because of concerns about the stability of the cavities used for waste placement, the German government decided to close the repository in 1998, and to leave the waste drums in place.

The Konrad Repository is a former iron ore mine located in Lower Saxony in North Germany (Berg *et al.*, 1987). The iron was mined from 1960 to 1976. The host rock is Upper Jurassic limestone. In 2002, the Konrad Mine received a license for its conversion into a geological repository for low-level and intermediate-level waste to be placed in chambers at a depth of 800–1,300 m. With a planned capacity of 303,000 m^3, the repository is currently under construction, and may be operational by 2027 (Bräuer, 2016).

A salt dome near the village of Gorleben in Lower Saxony has been investigated as a possible geological repository for all type of radioactive wastes: low-level, intermediate-level, high-level wastes, spent nuclear fuel, and decommissioning wastes. The site was selected in 1977, and studies began in 1979. The proposed disposal depth is from 840 to 1,200 m (Kaul and Rothemeyer, 1997). The geology and hydrogeology of the site have been studied via boreholes and core samples, seismic profiles, and groundwater sampling. The current timetable for the completion of the repository is not clear.

A somewhat different type of geological repository is in the Czech Republic. The former Czechoslovakia uses two chambers with a total volume of about 1,600 m^3 in a limestone mine dating from the 1940s near the village of Hostim in central Bohemia to dispose of approximately 400 m^3 of low- and intermediate level waste about 30 m below the surface. Operations lasted from 1959 to 1965, after which both chambers were backfilled with concrete. Total activity was less than 0.1 TBq (2 Ci). The repository was decommissioned in 1997 (Rempe, 2007). In 1974, Czechoslovakia also began to dispose of waste containing naturally occurring radionuclides in five chambers of an abandoned underground uranium mine near Jachymov. Almost 1,000 m^3 of space excavated in metamorphic rocks was originally available. The repository may continue to operate until about 2030 (Rempe, 2007).

The Czech Republic commenced underground disposal of radioactive waste in 1964 in the Richard Repository (Fig. 6.4), a former limestone mine, up to 70 m below the surface near Litomerice. A 5-m thick limestone bed — enclosed in a 50-m thick layer of impermeable marl — hosts the repository. Limestone extraction at the site had started in the mid-19th century. About 2,700 m^3 consisting of 25,000 waste packages containing 1,015 Bq (0.03 mCi) have

Figure 6.4. The Richard Repository for low-level radioactive waste in the Czech Republic (SÚRAO, 2020). Available at: https://www.surao.cz/en/ [Accessed 1 February 2020]. Used with permission of SÚRAO.

been filled, which leaves about 3,800 m^3 for future needs (Haverkamp *et al.*, 2005). The repository accepts institutional waste, derived from medical, industrial, research, and agricultural applications. Waste forms include radioactive sources and contaminated clothing, paper, and instruments. Projecting current trends into the future, the repository may continue operating until about 2070 (Rempe, 2007).

The National Radioactive Waste Repository in Bátaapáti, Hungary is used to store low-level, and intermediate-level radioactive wastes. Consisting of two inclined tunnels dug into granite, the wastes are buried at a depth of about 200–250 m below land surface. The first waste containers were placed in the repository in 2012.

Sweden has been using a geological repository for low- and intermediate-level radioactive waste for decades. Beginning in 1983, the Swedish Nuclear Fuel Waste Management Company (SKB) began construction of the Final Storage For Reactor Waste (SFR) which is uniquely located under the Baltic Sea near the municipality of Östhammar. It was excavated in a Pre-Cambrian granite gneiss with amphibolite and pegmatite dikes (Stille and Fredriksson, 1988). The central feature of the design of the facility is a 70-m tall, 30-m diameter concrete silo that is about 60 m below the top of the

Figure 6.5. The Swedish geological repository for low- and intermediate-level radioactive wastes (Final Storage For Reactor Waste) near Östhammar Sweden. Planned extension shown in the blue section (Bergström *et al.*, 2011). Used with permission of SKB.

crystalline host rock above (Fig. 6.5). In addition to the silo are four tunnels at depths of 50–129 m, providing a storage capacity of 63,000 m^3 of waste (Skogsberg and Ingvarsson, 2006). The maximum surface dose rate of waste containers in the silo is 500 mSv/h for remote-handled, intermediate wastes. One tunnel (BMA) is designated for intermediate wastes with a maximum surface dose of 100 mSv/h. The next tunnel is allocated for wastes yielding a maximum surface dose rate of 10 mSv/h. The last tunnel (BLA) is intended for low-level waste yielding $\leq$2 mSv/h. By 2006, 31,500 m^3 of waste were disposed, representing about 50% of its capacity. SKB is now planning to expand the facility (Fig. 6.5).

Finland has also been using geological repositories for low-, and intermediate-level radioactive wastes for decades. The construction of the repository at the Olkiluoto Power Plant began in 1988 and began operation in 1992. The construction of the repository at the Loviisa plant began in 1993, and waste emplacement began in 1998.

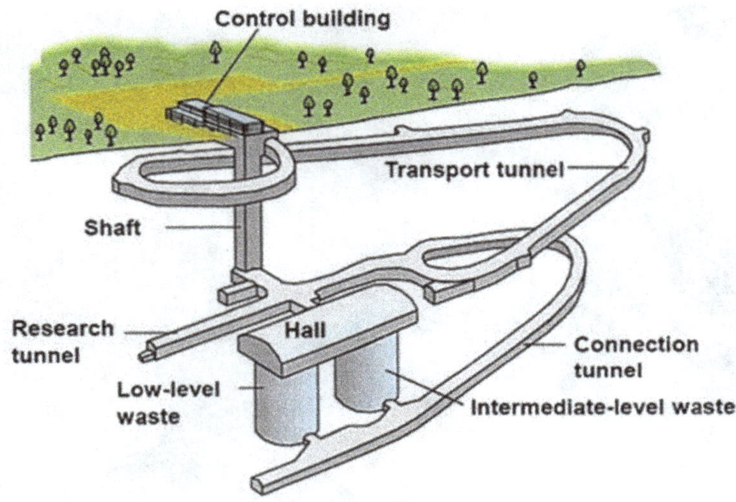

Figure 6.6. The geological repository for low- and intermediate-level radioactive waste at the Olkiluoto Power Plant, Finland (Posiva, 2020) Available at: http://www.posiva.fi/en [Accessed 1 February 2020]. Used with permission of Posiva Oy.

The disposal depth at Olikiluoto is from 60 to 100 m. The Olkiluoto site consists mainly of micaceous gneiss intercalated with sparsely fractured tonalite (Aikas and Anttila, 2008). The facility (Fig. 6.6) consists of two vertical silos: one for low-level radioactive wastes, and the other for intermediate-level wastes (Kekki and Tiitta, 2000). Both silos are lined with concrete. The type of containers used at Oilkiluoto include plastic sacks, 200-L drums, steel boxes, and concrete boxes. The Loviisa site consists entirely of Rapakivi granite, which is coarse-grained and porphyritic (Aikas and Anttila, 2008). The Loviisa repository consists of two tunnels for "dry maintenance waste" and "immobilized wet waste." In Olkiluoto, intermediate level ion-exchange resins are solidified in bitumen, packed into steel drums, and placed in the waste repository (Kekki and Tiitta, 2000). In Loviisa, the waste is solidified in cement and packed into concrete drums. In the future, the repositories will also be used for disposal of waste when the nuclear power plants are decommissioned (Aikas and Anttila, 2008).

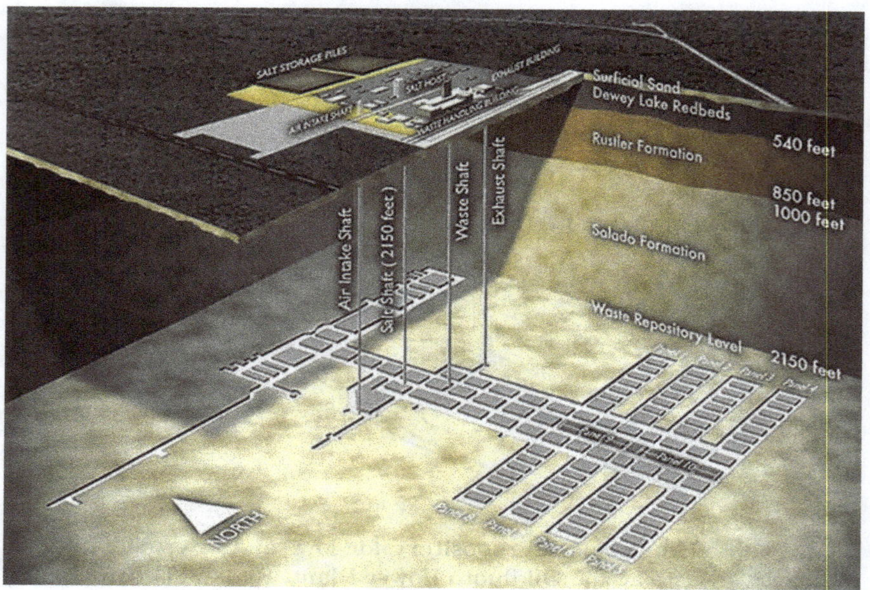

Figure 6.7. Simplified geological cross section and the Waste Isolation Pilot Plant in New Mexico (US DOE, 2020). Available at: https://www.energy.gov/em/ [2 February 2020].

In the US, the Waste Isolation Pilot Plant (WIPP) is a geological repository in New Mexico. Constructed and operated by the US DOE, it is a repository for the permanent disposal of transuranic (TRU) radioactive wastes in Permian-age salt deposits at a depth of 655 m below land surface (Fig. 6.7). Site characterization began in the mid-1970s, and disposal operations began in 1999 (Swift and Corbet, 2000). The Waste Isolation Pilot Plant Land Withdrawal Act of 1992 mandated that the capacity of the repository be 175,570 m^3, and that only TRU wastes from US DOE facilities are acceptable. Also, the Act mandated that the TRU wastes must have a surface dose rate of less than 1,000 rem/h.

The WIPP repository was constructed in the Salado Formation which is primarily halite ($NaCl$) with thinner layers of anhydrite ($CaSO_4$). Other minerals such as sylvite (KCl) and polyhalite ($K_2Ca_2Mg(SO_4)_4 \cdot 2H_2O$) have been observed (Lambert, 1992). The bedded salt of the Salado Formation is the primary barrier to radionuclide migration from the WIPP. The hydraulic conductivity of the Salado Formation is limited (10^{-14} to 10^{-9} cm/s), and the brine content is estimated to be 1–2% by weight (Swift and Corbet, 2000). Most of the sedimentary rocks at the WIPP were deposited in the Permian Period during the subsidence of the Delaware Basin. Increasing isolation of the marine basin from the open ocean resulted in the deposition of evaporite minerals. As of 2013, 87,369 m^3 of TRU waste have been disposed at the WIPP, filling about 50% of its mandated capacity (US DOE, 2020). WIPP was designed to dispose of two kinds of TRU waste. Contact-Handled is defined as having a surface dose that is less than 200 mrem/h. Remote-Handled TRU waste can have a dose rate from 200 to less than 1,000 rem/h. About 96% of the waste to be disposed of at WIPP is Contact-Handled TRU waste. Contact Handled TRU waste barrels and boxes are stacked in rows on the floor of WIPP's underground disposal rooms, while Remote Handled TRU waste canisters are placed in boreholes drilled into the walls of the same rooms.

In Canada, the Nuclear Waste Management Organization has proposed the development of a Deep Geologic Repository (DGR) for low- and intermediate-level wastes near the Western Waste Management Facility near Tiverton, Ontario (Fig. 6.8). The proposed repository would have been at a depth of about 680 m below ground surface with the Ordovician-age Cobourg Formation, a 200-m thick argillaceous limestone (Gierszewski, 2008). The disposal capacity of the repository was about 200,000 m^3 and would have been be available to only nuclear power plants in Ontario.

The horizontal hydraulic conductivity of the limestone for the proposed Canadian repository has been estimated as 1×10^{-12} cm/s, and the groundwater — a saline brine with a total dissolved content of 230–270 g/L — is stagnant (Raven *et al.*, 2010). Therefore, potential solute transport from the repository would be diffusion-limited. Raven *et al.* (2010) summarized that the vertical diffusion coefficient for anionic iodide, as measured using rock core samples collected during site characterization, were on the order of 3×10^{-9} cm^2/s. The Environmental Impact Assessment of the proposed repository was under review by the Minister of Environment and Climate Change. In 2020, some members of the Saugeen Ojibway Nation objected to

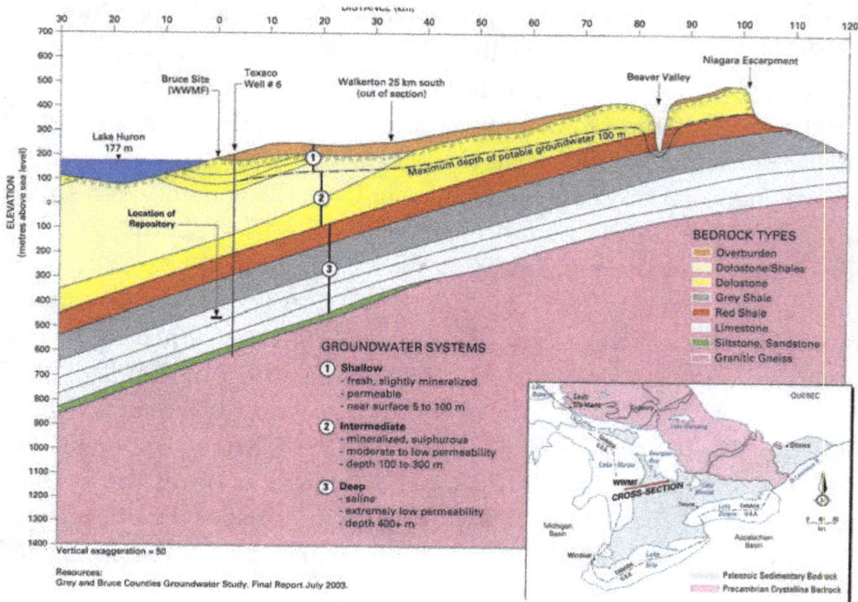

Figure 6.8. Geological cross section showing the location of the proposed Deep Geological Repository (DGR) for low- and intermediate-level radioactive wastes near Tiverton, Ontario, Canada. Used with permission of Ontario Power Generation.

the proposed repository because of concerns that it would somehow result in contamination of Lake Huron. Ontario Power Generation cancelled the project and plans to consider other potential sites.

The Russian Federation has been using geological repositories for liquid radioactive wastes since 1962. Radioactive liquids have been injected into relatively water-permeable geological formations using a process called deep-well injection. The Soviet Union injected about 50 million m^3 of radioactive waste, much of it being reprocessing wastes (Rempe, 2007). As discussed by Compton *et al.* (2000), Krasnoyarsk-26 was an underground facility in Western Siberia that produced plutonium for weapons during the Cold War by reprocessing spent nuclear fuel. The liquid wastes are transported to the Severny Site, an injection facility with 12 injection wells, and numerous observation wells. There have been three types of waste streams: (1) a high-level, acidic (pH 1–3) solution, (2) an

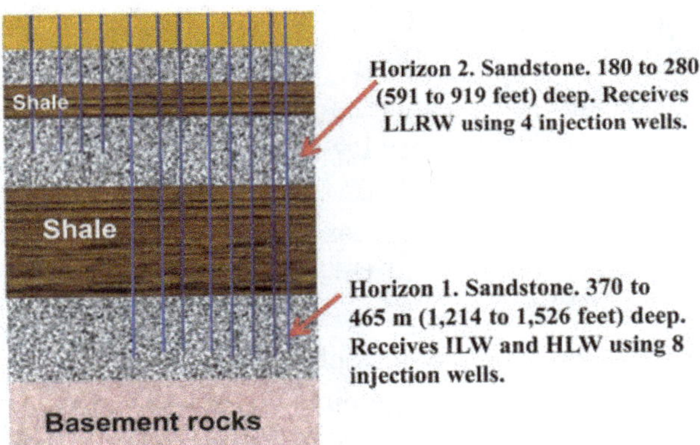

Figure 6.9. Generalized geological column for the Severty Site, western Siberia, Russian Federation.

intermediate-level alkaline waste derived from neutralizing the acidic raffinate, and (3) a low-level waste solution. The Severny Site consists of two relatively water-permeable, Jurassic-age sandstones that are stratigraphically below relatively impermeable shales (Fig. 6.9). The lower most sandstone layer, called the Horizon I occurs at a depth of 370–465 m (1,214–1,526 feet). Eight injection wells are used to pump the intermediate-level, and the high-level wastes into the deeper formation. Horizon II is also a sandstone layer at 180–280 m (591–919 feet) depth. The low-level waste is injected into this upper sandstone using four wells. The migration and chemical fate of the injected radionuclides was modeled by Compton *et al.* (2000) and Rumynin *et al.* (2005). Rybalchenko *et al.* (2005) concluded that deep-well injection of radioactive wastes in Russia will likely continue for the foreseeable future.

6.4 Repositories for Spent Nuclear Fuel and High-Level Wastes

6.4.1 *The US Yucca Mountain Nuclear Waste Repository*

The Yucca Mountain Nuclear Waste Repository may be the first geological repository in the US for commercial spent nuclear fuel, legacy high-level waste from US DOE, and US Navy spent nuclear fuel. Located in Nevada, Yucca Mountain is a ridge composed of layers of volcanic tuff which is consolidated volcanic ash that accumulated from volcanic activity during the Miocene epoch. The layers are tilted between 5° and 10° to the east (Rechard *et al.*, 2013). The site is in a semi-arid environment that receives an annual rainfall of less than about 15 cm, and of this amount about 95% evaporates. Only about 1–2% infiltrates into the ground (Eckhardt, 2000). In addition to its location in a dry climate, the proposed repository is in the unsaturated (vadose) zone which extends to a depth of about 540–600 m below the surface. The proposed disposal area is about 300 m below the surface (Fig. 6.10). US DOE began studies of Yucca Mountain in 1978. The main tunnel of the Exploratory Studies.

Facility is a u-shaped excavation that is 7.6 m wide and 8.1 km long. Another 3.2-km tunnel (Cross Drift) branches off the main

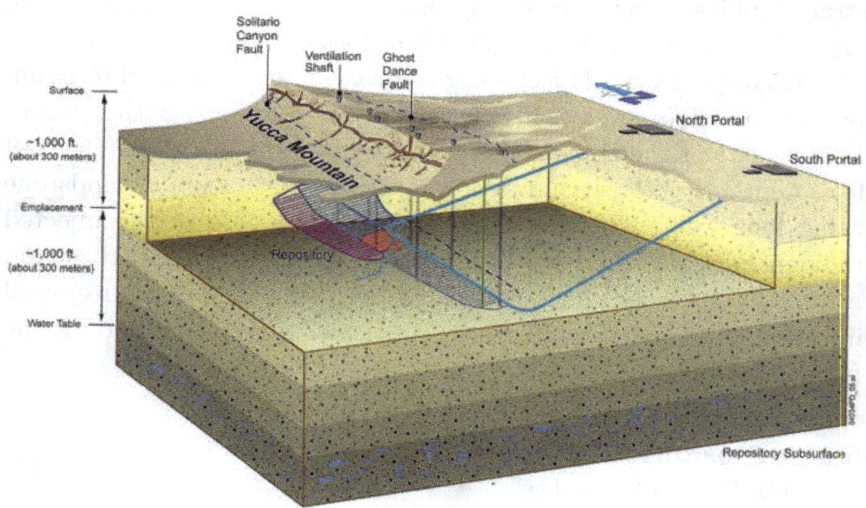

Figure 6.10. The proposed Yucca Mountain Nuclear Waste Repository in Nevada, US (US EPA, 2019. Available at: https://www.epa.gov/radiation/what-yucca-mountain-repository [Accessed 2 February 2020].

tunnel. The Nuclear Waste Policy Act of 1982 set a statutory limit for the repository at 70,000 metric tonnes of heavy metal,[1] and that 10% (7,000 tonnes) would be allocated to radioactive wastes from US DOE-managed nuclear materials.

The layer planned for disposal is the Topopah Spring Tuff, which is a quartz latite-rhyolite that reaches a maximum thickness of 350 m and is relatively homogeneous in terms of mineralogy and chemical composition (Peterman and Cloke, 2002). In the lower part of the welded Topopah Spring Tuff, however, there is a zone of smectite clay (a hydrous aluminum phyllosilicate) and zeolite minerals that formed by the alteration of the volcanic ash in the presence of alkaline groundwater. Vaniman *et al.* (2001) presented a detailed analysis of the mineralogical composition of core samples collected from the Topopah Spring Tuff and the non-welded Calico Hills Formation below it (Table 6.2). There are other intervals with concentrated and dispersed zeolites that are below the zone of saturation. As shown in Table 6.2, the amount of smectite was variable, but can be as much as 8% by weight in the lower Topopah Spring Tuff. The distribution of the zeolite clinoptilolite ($(Na, K, Ca)_{2-3}Al_3(Al, Si)_2Si_{13}O_{36} \cdot 12H_2O$)

Table 6.2. Partial mineralogical composition (weight-percent) of core samples collected from the volcanic tuffs below the proposed geological repository at Yucca Mountain, Nevada (from Vaniman *et al.*, 2001).

Mineral	Lower Topopah Spring Tuff (44 samples)	Calico Hills Formation (41 samples)
Smectite	Trace to 8	0–11
Clinoptilolite	0–80	0–71
Mordenite	0–3	0–20
Amorphous silica	0–96	0–90

[1]The term heavy metal in this application is a unit of mass used to quantify uranium, plutonium, thorium, and mixtures of these relatively heavy and dense metals.

varies considerably, but was a major component in some samples. Mordenite ($(Ca, Na_2, K_2)Al_2Si_{10}O_{24} \cdot 7H_2O$), another zeolite, was less detected in the samples, but was more abundant in the Calico Hills Formation. These alteration products have the potential to retain dissolved radionuclides primarily by ion exchange. Hence, they can serve as natural barriers to the migration of radionuclides that may be mobilized from the proposed repository. For example, Bertetti *et al.* (1998) measured the sorption of neptunium (V) by clinoptilolite and found that the extent of sorption increased with the pH of the liquid phase. Pabalan *et al.* (1998) observed that the extent of uranium (VI) sorption by clinoptilolite was greatest at about pH 6. As a background for the potential efficacy of zeolites at Yucca Mountain, Misaelides (2011) provided a review of the application of natural and modified zeolites in environmental applications.

The feasibility of the Yucca Mountain site as a geological repository was characterized by numerous performance assessments to determine potential environmental impacts, repository design options, an assessment for the US Congress, and additional updates and analyses for the license application to construct the repository. In attempt to document the sequence of events, Rechard (2013) summarized the progression of four early performance assessments conducted between 1982 and 1995. Rechard *et al.* (2013) summarized the investigations conducted to characterize the geological barriers available at Yucca Mountain. Rechard and Stockman (2013) summarized the results of performance assessments conducted between 1984 and 2008 on the modeling of waste container degradation and the subsequent mobilization of radionuclides. Hansen *et al.* (2013) summarized the results of the "total system model" in which the likelihood and potential impacts of five events on the operation of the repository were evaluated. These events were: (1) early waste package failure, (2) early drip-shield failure, (3) igneous activity, (4) seismic activity, and (5) seismic-fault activity.

In 2006, the US Senate Committee on Environmental and Public Works issued a white paper in support of the proposed repository (US Senate, 2006). In 2008, US DOE submitted a license application to the US NRC, seeking authorization to construct the repository with the goal of being available to receive wastes in 2017. However, because of strong political opposition from legislators in Nevada, the Obama Administration eliminated all funding for the Yucca Mountain program. In 2020, because of same political opposition, the Trump Administration also decided against providing funds for the constructing the repository.

Not in a Million Years

The US EPA required that the US DOE consider the long-term impacts of early waste package failure, early drip-shield failure, igneous activity, seismic activity, seismic-fault activity, and climate change during a one million-year period after closure of the proposed Yucca Mountain Nuclear Waste Repository. In 2008, the end results of numerous Performance Assessments are that the repository must provide a dose limit of 0.15 mS/year (15 mrem/year) for the first 10,000 years after disposal for each nearby resident of Yucca Mountain. The repository is then required to limit doses to 1 mSv/year (100 mrem/year) from 10,000 to 1,000,000 years. As summarized by Swift *et al.* (2014), it was envisioned in the Performance Assessments that from 100 to 1,000 years, the waste containers would remain intact and that the radioactive decay heat from the canisters would evaporate the groundwater near the canisters. From 1,000 to 10,000 years, it was assumed that some of the canisters would begin to corrode. The heat of decay has gradually decreased, and groundwater enters the repository. Some of the engineered barriers begin to degrade, and there is a minor release of radionuclides. The radiation doses are dominated by technetium-99 and iodine-129. From 10,000 to 1,000,000 years after closure, a significant number of the waste canisters and the titanium dip shields fail because of corrosion. Groundwater dilutes and transports radionuclides

(*Continued*)

(*Continued*)

from the repository. The radioactive doses are dominated by neptunium-137 but are still less than 1 mSv/year — according to numerous model simulations. For comparisons, the half-life of neptunium 137 is 2.144×10^6 years. As many have, the reader may find it inconceivable that a performance standard for a geological repository has been evaluated for such a long period. Consider that — one million years ago — *homo sapiens* had not yet evolved on the earth. It is difficult to predict who or what will be protected from radiation emitted by spent nuclear fuel in the far, distant future.

6.4.2 *The Kingdom of Sweden*

Sweden has the distinction for establishing a central interim geological repository for the wet storage of spent nuclear fuel for 30–40 years while preparing a deeper repository for final disposal. Located near Oskarshamn in southeast Sweden, the below-ground portion of the facility called Clab (Central Holding Storage for Spent Nuclear Fuel) consists of two tunnels, excavated in granite, each containing five concrete pools. Each tunnel is 120 m long and is about 30 m below ground surface. The repository began receiving spent nuclear fuel in 1985 and has a capacity of 8,000 tonnes of spent nuclear fuel. In 2019, Clab contained about 6,700 tonnes of spent nuclear fuel (Swedish Nuclear Fuel and Waste Management Company, 2020. Available at: http://www.skb.se [Accessed 2 February 2020].

A siting process was begun in about 1993 to locate a potential repository for the final disposal of spent nuclear fuel removed from cooling at Clab. In 2009, a site near the Forsmark Power Plant in Osthammar was selected on the basis of having the best geological conditions, and the application for construction was sent to the Swedish Radiation Safety Authority and the Land and Environmental Court in Stockholm in 2011. The application is currently under review. If the review process moves forward, the repository may be operational by about 2023. The capacity of the proposed repository

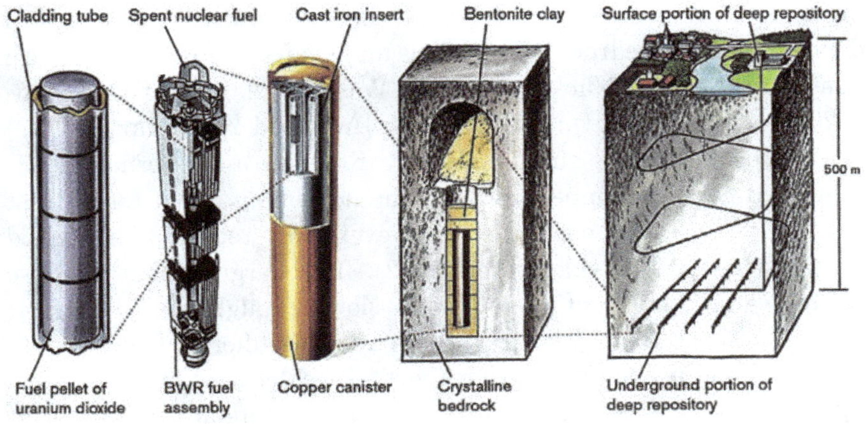

Figure 6.11. The KBS-3 Method in which spent nuclear fuel is first placed in a copper canister which is inserted in bedrock and sealed with bentonite clay in a geological repository (Swedish Nuclear Fuel and Waste Management Company, 2020. Available at: http://www.skb.se [Accessed 2 February 2020].

is 12,000 tonnes of spent fuel, and the wastes will be disposed in 1.9 billion-year-old granite of the Fennoscandia Bedrock Shield. A major component of the waste management program in Sweden is the KBS-3 Method (Fig. 6.11). In the Nuclear Activities Act of 1984, the Swedish Government advocated the use of the KBS-3 Method. KBS is derived from Kärn Bränsle Säkerhet, or "nuclear fuel safety" in Swedish. The principle of the KBS-3 Method is to place the fuel assemblies into a boron-steel canister, then encase it in a copper canister that is 50 mm thick. Each capsule would be placed in its own hole in the repository then packed bentonite clay about 500 m down into the bedrock. As discussed by Rosborg and Werme (2008), the KBS-3 Method was based on natural analogues of geological repositories. Copper was chosen based on the long-term stability of natural deposits and archaeological artifacts as analogues of waste packages. It is estimated that the tunnels will be about 250 m long, and that they will be positioned about 40 m apart. In the base of the tunnels there will be disposal holes that are about 6 m apart. The copper canisters will be placed in the disposal holes and embedded in the bentonite-clay buffer. When all the spent nuclear fuel has been deposited in the bedrock, the tunnels and shafts will be filled in with bentonite (Swedish Nuclear Fuel and Waste Management Company, 2020. Available at: http://www.skb.se [Accessed 2 February 2020].

Corrosion studies summarized by Rosborg and Werme (2008) suggested that the copper canisters should remain intact for at least 100,000 years. This conclusion, however, was recently challenged by Szákalos and Seetharamam (2012). They argued that because groundwater movement is relatively slow through the granite, it may take about 6,000 years for the bentonite buffer to become fully saturated and pressurized with respect to the depth of burial. During this re-saturation period, Szakalos and Seetharamam proposed that the copper cylinders would be exposed to corrosive agents that had not been taken into account during the earlier assessments. They proposed that hydrogen sulfide, methane, and hydrogen gases may degas from the slowly moving groundwater and react with the cylinders. Moreover, Szakalos and Seetharamam argued that chlorides from groundwater evaporating because of the decay heat could contribute to corrosion. In brief, Szakalos and Seetharamam suggested that corrosion during the first 1,000 year after emplacement may be more problematic than SKB anticipates. The long-term stability and design of the copper canisters is currently under review.

6.4.3 *Republic of Finland*

A geological repository for spent nuclear fuel in Finland is under construction at Olkilutoto. Site selection began in 1983. Posiva Oy was established in 1995 as a company owned by the two major nuclear power companies to dispose of spent nuclear fuel in Finland. In 1999, Posiva Oy submitted an application for construction of the repository to the Finish Parliament where it was approved, and construction began in 2004. The facility, named "Onkalo," is being built in four phases, the first two being focused on downward excavation and bedrock characterization. The last two phases will focus on the actual repository excavation such that disposal can begin in 2020s (Ikonen *et al.*, 2006), making it the first geological repository for spent

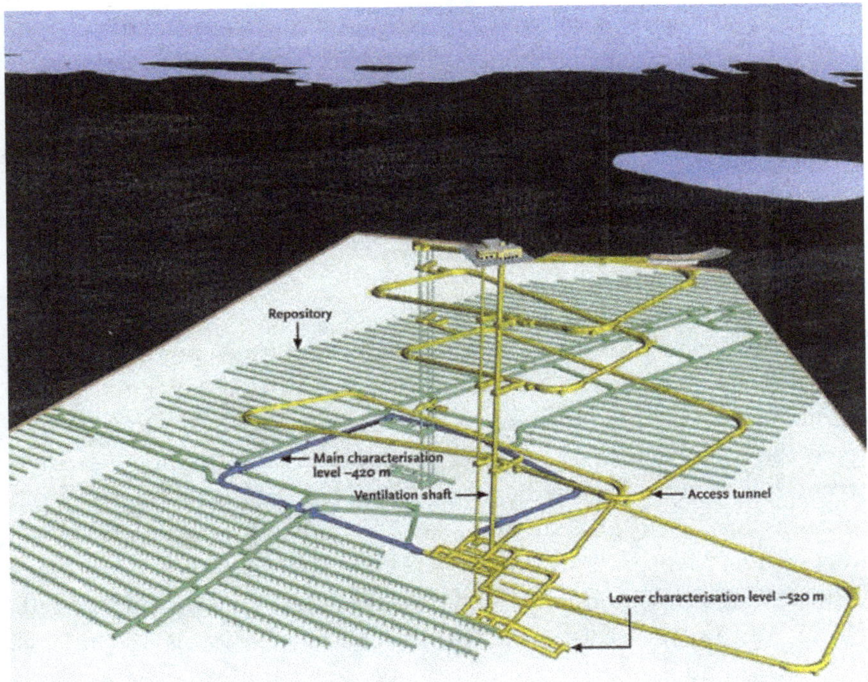

Figure 6.12. The Onkalo repository for spent nuclear fuel in Olkiluoto, Finland (Posiva Oy, 2017). Used with permission of Posiva Oy.

nuclear fuel in the world (Fig. 6.12). The depth of the repository will be about 520 m (1,710 feet). Ikonen *et al.* (2006) the management of groundwater leakage in the tunnels, via fractures, required extensive grouting. Once disposal commences, Posiva Oy plans to use the KBS-3 Method to isolate the spent fuel canisters. The capacity of the repository is 9,000 tonnes of uranium.

Posiva Oy has published numerous reports about Onkalo on site geology, impacts of future climate change, solute transport, waste package stability and other interesting aspects. Posiva Oy (2012) concluded that the bedrock at Olkiluoto, the Pre-Cambrian Fennoscandian Shield, consists of mainly gneisses and pegmatitic granites and diabase dykes. The bedrock has been subjected to extensive hydrothermal processes, yielding alteration minerals (illite, kaolinite, sulfides and calcite). Groundwater moves through fractures and brittle deformation zones. The pore water within the rock matrix is stagnant. At the repository depth, the groundwater is saline — having a total dissolved solids content of about 12 g/L. The groundwater is also chemically reduced except in areas of shallow groundwater recharge.

6.4.4 *French Republic*

The 1991 Waste Management Act established the Agence Nationale pour las gestion des Déchet Radioactits (ANDRA) as the National Radioactive Waste Management Agency. In 1999, ANDRA authorized the construction of an underground research laboratory at Bure to conduct studies for the disposal of vitrified high-level wastes (from reprocessing used nuclear fuel) and long-lived intermediate-level wastes. The Meuse/Haute-Marne Underground Research Laboratory is in the Jurassic-age Callovo-Oxfordian formation at a depth of 420–550 m. Within the eastern Paris Basin, the formation is 130 m thick (Fig. 6.13) and is relatively impermeable to groundwater — having a hydraulic conductivity of 10^{-12} to 10^{-10} cm/s — and a water content of about 8% by weight (Vinsot *et al.*, 2008).

The Callovo-Oxfordian formation is a sequence of clay-rich limestones, and silty and sometimes calcareous shales. The lower part of the formation is dominated by illite and kaolinite, with sporadic occurrences of calcite and pyrite (Descostes *et al.*, 2008). The Callovo-Oxfordian sequence has been studied extensively by ANDRA. For example, because of the relatively impermeable portion of the shales, Descostes *et al.* (2008) measured the rate of diffusion of four radioactive anionic tracers ($^{36}Cl^-$, $^{125}I^-$, $^{35}SO_4^{2-}$, and $^{75}SeO_3^{2-}$) under laboratory conditions using core samples collected from the Callovo-Oxfordian argillites and the Oxfordian limestones that are stratigraphically above. The results of such studies provide data on how rapidly radionuclides could move by diffusion alone through the shale barrier.

Because of current French regulations, the Underground Research Laboratory cannot be used to dispose of radioactive wastes in

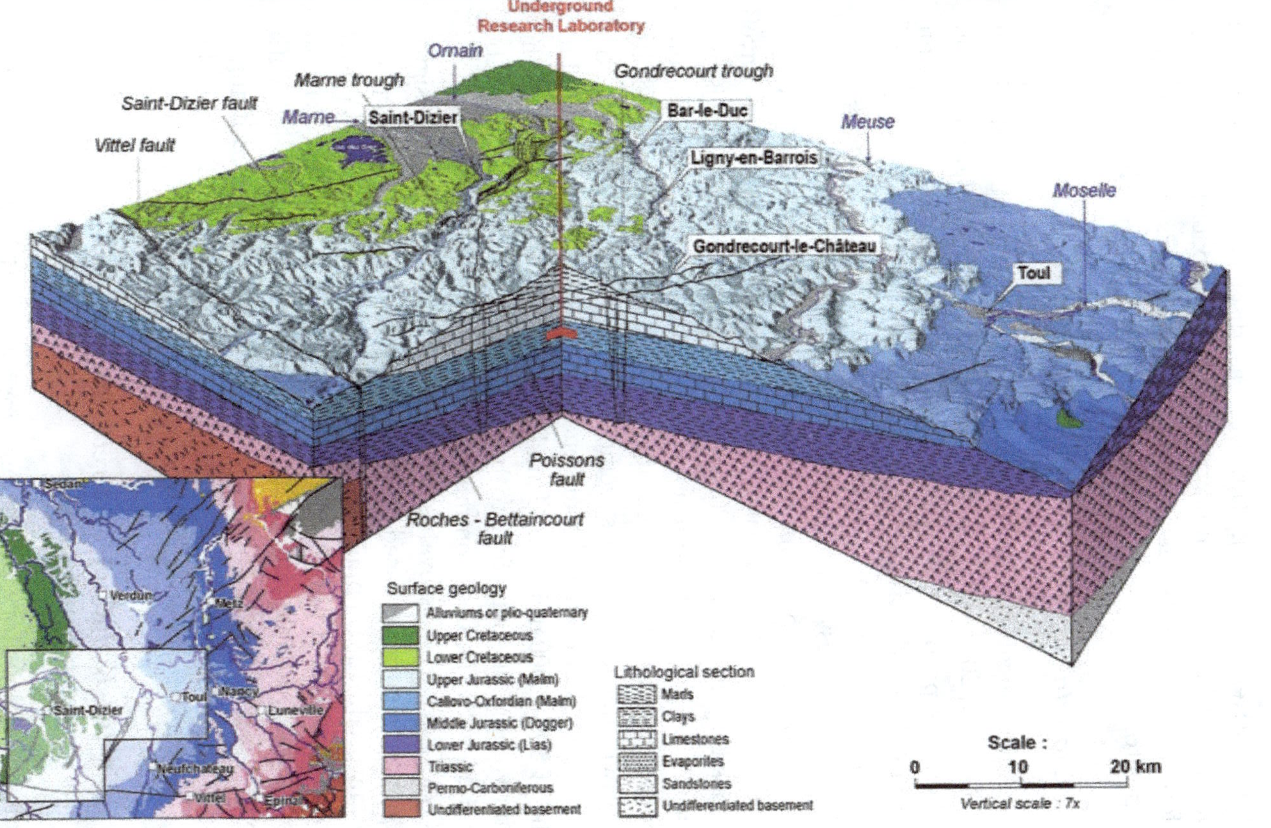

Figure 6.13. Geological block diagram of the Meuse/Haute Marne Underground Research Laboratory (Linard *et al.*, 2011).

the future (Rempe, 2007). However, as discussed by Dupis and Ouzounian (2011), the feasibility of the Callovo-Oxfordian formation to safely contain wastes has been demonstrated. It is envisioned that the repository will be operational by about 2025 (see Chapter 10).

6.4.5 *Canada*

A deep, geological repository was first advocated in Canada in 1974 (Ramana, 2013). The Nuclear Waste Management Organization (NWMO) has proposed to site and construct a repository using an approach called Adaptive Phased Management. This approach calls for the spent fuel to be retrievable, and that the siting process must identify a willing host community. NWMO (2012) envisioned that site selection and regulatory approval process would take about 15 years, followed by a 10-year construction period. NWMO expects to have the repository available by 2035 (WNA, 2019). NWMO envisions that the repository will be a network of tunnels, access drifts, and placement rooms, all excavated at a single depth of 500 m. The site geology will depend on the specific community that agrees to host the repository. An earlier recommendation was to bury spent fuel at depths of 500–1,000 m in the Canadian Shield (Ramana, 2013). The specific volume of spent fuel to be placed in the repository will be negotiated with the host community.

NWMO is currently evaluating different waste container designs, but a copper container based on the Swedish KBS-3 Method appears to be favored. In keeping with the Swedish approach, NWMO proposed that that copper containers will be placed on either vertical or horizontal boreholes in the placement rooms, then sealed with bentonite. NWMO is currently in the early stages of identifying a willing host community for the repository (Ramana, 2013), with the goal of beginning construction in about 2026.

6.4.6 *United Kingdom of Great Britain and Northern Ireland*

A geological repository is planned for radioactive wastes in the United Kingdom (UK), but like Canada, the UK is currently in the early stages of identifying a willing host community. Chapman and Hooper (2012) suggested that the Government of UK plans to dispose intermediate-level wastes in an engineered geological repository by about 2040. Spent fuel and high-level reprocessing wastes may be disposed at a later date (see also Chapter 10).

6.4.7 *The Deep-Borehole Repository*

An alternative to the deep geological repositories described above is the deep-borehole concept. Designs for the burial of radioactive wastes in boreholes was first proposed in about 1965, but this concept has since evolved for the disposal of spent nuclear fuel and vitrified high-level waste. Gibb (2000) proposed a drill hole with a diameter of 0.5 m be drilled to depth of 4–5 km such that the bottom of the borehole would reach into Pre-Cambrian granitic rocks (Fig. 6.14). Once the waste containers are stacked, the borehole would be

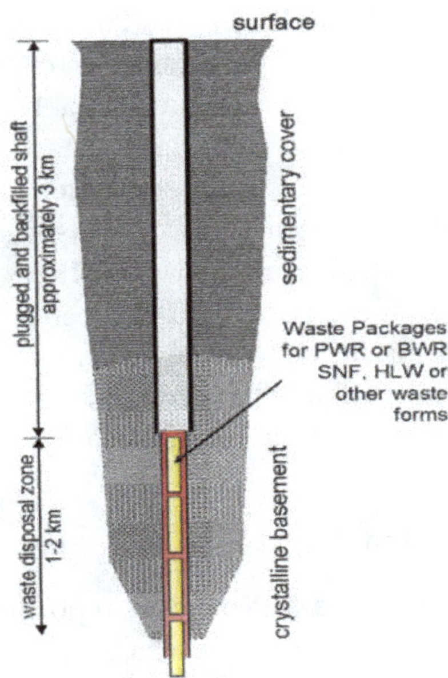

Figure 6.14. Deep borehole disposal of spent nuclear fuel (Brady *et al.*, 2009).

backfilled with bentonite and crushed granite. Gibb envisioned that the decay heat from the waste package could partially melt the backfill and rock surrounding the waste containers. When cooling conditions develop, the melted rocks would recrystallize, forming a seal about the waste-disposal interval. Gibb *et al.* (2008) also suggested that deep borehole disposal would be a logical choice for excess plutonium that is not destined for mixed oxide fuel fabrication, thus eliminating proliferation issues.

Deep borehole disposal has some advantages over conventional repositories. Granitic basement rocks are common in the US at depths of 2,000–5,000 m, and therefore the site-selection process for potential boreholes would not be as location dependent. Moreover, groundwater movement would be of little concern at such depths. The number of wells could be increased as the need for greater capacity demands. There are two disadvantages: the wastes would not be easily retrievable, and the physical size of the borehole would limit the size of the waste container that could be lowered down to the disposal interval. Brady *et al.* (2009) presented a detailed assessment of the feasibility of deep borehole disposal in the United States. Before this disposal technique could be considered in the US, however, the Nuclear Waste Policy Act would have to be amended because it mandates that the Yucca Mountain Nuclear Waste Repository be first licensed for the disposal of spent nuclear fuel. Geringer *et al.* (2013) presented a feasibility analysis for deep borehole disposal for spent nuclear fuel in central Illinois. The role of deep borehole disposal in the management of radioactive wastes remains as a potential option. The reader may also find IAEA (2009) and Ketner (2014) useful sources of information.

6.5 Review Questions

1. What is the purpose of a geological repository for radioactive wastes? What are the desirable characteristics of a geological repository?
2. What is retrievable storage? What are the benefits and potential problems with this concept?

3. Describe how natural analogues can provide insight about the potential efficacy of geological repositories.
4. Summarize efforts in Germany and the Czech Republic to use former mines as geological repositories.
5. Describe the Waste Isolation Pilot Plant as a geological repository.
6. Make a table that contrasts and compares the geological conditions, facility design, and intended wastes to be disposed in Finland, Sweden, Canada, and France.
7. Compare deep-well injection with deep borehole disposal of radioactive wastes.
8. Summarize the design planned for disposing radioactive wastes at the proposed Yucca Mountain Nuclear Waste Repository.
9. Describe Sweden's KBS-3 Method for isolating spent nuclear fuel.
10. What do Canada, Sweden, and Finland have in common with respect to the intended waste package for spent nuclear fuel?

Bibliography

Aikas, T. and Anttila, P. (2008). Repositories for Low- and Intermediate-Level Radioactive Wastes in Finland. *Reviews in Engineering Geology*, 19, pp. 67–71.

Allan, C. J. and Nuttall, K. (1997). How to Cope with the Hazards of Nuclear Fuel Waste. *Nuclear Engineering and Design*, 176, pp. 51–66.

Berg, H. P., Brennecke, P. W., and Thomauske, B. R. (1987). The German Konrad Repository project. *Progress in Nuclear Energy*, 20, pp. 255–307.

Bergström, U., Pers, K., and Almén, Y. (2011). *International Perspective on Repositories for Low Level Waste.* Swedish Nuclear Fuel and Waste Management Company (Report number SKB R-11-16).

Bertetti, F. P., Pabalan, R. T., and Almendarez, M. G. (1998). Studies of Neptunium^{V} Sorption on Quartz, Clinoptilolite, Montmorillonite, and α-Alumina. In Jenne, E. A. (ed.), *Adsorption of Metals by Geomedia*, Chapter 4. Academic Press, San Diego, California, pp. 131–148.

Berzero, A. and D'Alessandro, M. (1990). *The Oklo Phenomenon as an Analogue of Radioactive Waste Disposal.* Commission of the European Communities, Nuclear Science and Technology (Report EUR 12941 EN). Bjursrom, S. 1989. Storage of Nuclear Waste in Sweden. *Tunneling and Underground Space Technology*, 4, pp. 139–142.

Brady, P. V., Arnold, B. W., Freeze, G. A., Swift, P. N., Bauer, S. J., Kanney, J. L., Rechard, P. R., and Stein, J. S. (2009). *Deep Borehole Disposal of*

High-Level Radioactive Waste. Sandia National Laboratories (Sandia Report, SAND2009-4401).

Bräuer, V. (2016). Current Status of Nuclear Waste Disposal in Germany. In Faybishenko, B., Birkholzer, J., Sassani, D. and Swift, P. (eds.), *International Approaches for Deep Geological Disposal of Nuclear Waste: Geological Challenges in Radioactive Waste Isolation. Fifth Worldwide Review.* Chapter 9, Lawrence Berkeley National Laboratory (Report number LBNL-1006984) pp. 9–16.

Brookins, D. G. (1984). *Geochemical Aspects of Radioactive Waste Disposal.* Springer-Verlag, New York.

Brookins, D. G. (1990). Radionuclide Behavior at the Oklo Nuclear Reactor, Gabon. *Waste Management*, 10, pp. 285–296.

Cameco. (2012). *Cigar Lake Project. Northern Saskatchewan* (NI 43-101 Technical Report).

Chapman, N. and Hooper, A. (2012). The Disposal of Radioactive Waste Underground. *Proceedings of the Geologists' Association*, 123, pp. 46–63.

Chapman, N. A. and McKinley, I. G. (1987). *The Geology of Disposal of Nuclear Waste.* John Wiley and Sons, U.K.

Compton, K. L., Novikov, V., and Parker, F. L. (2000). *Deep Well Injection of Liquid Radioactive Waste at Krasnoyarsk-26: Volume I.* International Institute for Applied Analysis, Laxenburg, Austria (Report RR-00-1).

Cowan, G. A. (1976). A Natural Fission Reactor. *Scientific American*, 235, pp. 36–47.

Cramer, J. J. (1995). Cigar Lake: A Natural Example Of Long-Term Isolation Of Uranium. *Radwaste*, May 1995, pp. 36–40.

Curtis, D., Benjamin, T., DeLaeter, J., Delmore, J. E., and Maeck, W. J. (1989). Fisson Product Retention in the Oklo Natural Fission Reactors. *Applied Geochemistry*, 4, pp. 49–62.

Descostes, M., Blin, V., Bazer-Bachi, F., Meier, P.,Grenut, B., Radwan, J., Schlegel, M. L., Buschaert, S. Coelho, D., and Tevissen, E. (2008). Diffusion Of Anionic Species in Callovo-Oxfordian Argillites and Oxfordian Limestones (Meuse/Haute-Marne, France). *Applied Geochemistry*, 23, pp. 655–677.

Dupuis, M. C. and Ouzounian, G. (2011). The French geological repository project; a converging approach. *Waste Management Conference, Phoenix, Arizona, Feb. 27-March 3, 2011.*

Eckhardt, R. C. (2000). *Yucca Mountain. Looking Ten Thousand Years into the Future.* Los Alamos National Laboratory (Science Number 26), pp. 464–488.

Geringer, R. J., Zhou, J., and Mouche, P. (2013). Deep Borehole Storage of Nuclear Waste In Central Illinois Geology: A Technical, Economic, and Political Feasibility Analysis. In *American Nuclear Society 2013 Student Conference, Massachusetts Institute of Technology*, Boston MA, April 4–6, 2013.

Gibb, F. G. F. (2000). A New Scheme for the Very Deep Geological Disposal of High-Level Radioactive Waste. *Journal of the Geological Society, London*, 157, pp. 27–36.

Gibb, F. G. F., Taylor, K. T., and Burakov, B. (2008). Deep Borehole Disposal of Plutonium. In *Materials Research Society Symposium Proceedings*, 1107. DOI: https://doi.org/10.1557/PROC-1107-51.

Gierszewski, P. (2008). Overview of Ontario Power Generation's proposed Deep Geologic Repository for Low and Intermediate Level Waste at the Bruce Site, Ontario, Canada. In *Waste Management Conference*, Phoenix, Arizona, Feb. 24–28, 2008.

Hansen, C. W., Birkholzer, J. T., Blink, J., Bryan, C. R., Chen, Y., Gross, M. B., Hardin, E., Perry, F. V., Houseworth, J., Howard, R., Jarek, R., Lee, K. P., Lester, B., Martin, P., Mattie, P. D., Mehta, S., Robinson, B., Sassani, D., Sevougian, S. D., Stein, J. S., and Wasiolek, M. (2014). Overview of Total System Model Used for the 2008 Performance Assessment for the Proposed High-Level Radioactive Waste Repository at Yucca Mountain, Nevada. *Reliability Engineering and System Safety*, 122, pp. 249–266.

Haverkamp, B., Biurrun, E., and Kucerka, M. (2005). Update of the Safety Assessment Of The Underground Richard Repository, Litomerice. In *Waste Management Conference*, Tucson, Arizona, February 27 to March 3, 2005.

IAEA (2009). *Borehole Disposal Facilities for Radioactive Waste.* International Atomic Energy Agency, Specific Safety Guide (Report No. SSG-1).

Ikonen, A., Yla-Mella, M., and Aikas, T. (2006). Underground Characterization and Research Facility Onkalo. *European Nuclear Society*, Topseal 2006, Olkiluoto, Finland, September 17–20, 2006.

Kaul, A. and Rothemeyer, H. (1997). Investigation and evaluation of the Gorleben site: a status report. *Nuclear Engineering and Design*, 176, pp. 83–88.

Kekki, T and Tiitta, A. (2000). *Evaluation of the Radioactive Waste Characterization At Olkiluoto Nuclear Power Plant.* Radiation and Nuclear Safety Authority (Finland) (Report STUK-YTO-TR 162, Helsinki, Finland.

Ketner, J. (2104). *Deep Borehole Disposal of Radioactive Waste and Spent Nuclear Fuel.* Nova Science Publishers, Inc., New York.

Lambert, S. J. (1992). Geochemistry of the Waste Isolation Pilot Plant (WIPP) Site, Southeastern New Mexico, U.S.A. *Applied Geochemistry*, 7, pp. 513–531.

Linard, Y., Vinsot, A., Vincent, B., Delay, J., Wechnew, S., Vaissiere, R. D. L., Scholz, E., Garry, B., Lundy, M., Dewonck, S., and Vigeron, G. (2011). Water flow in the Oxfordian and Dogger limestone around the Meuse/Haute-Marne Underground Research Laboratory. Physics and Chemistry of the Earth, Parts A/B/C Volume 36, Issues 17–18, pp. 1450–1468.

Min, M. Z., Zhai, J. P., and Fang, C. Q. (1998). Uranium-Series Radionuclide and Element Migration Around The Sanerliu Granite-Hosted Uranium Deposit In Southern China As A Natural Analogue For High-Level Radwaste Repositories. *Chemical Geology*, 144, pp. 313–328.

Misaelides, P. (2011). Application of natural zeolites in environmental remediation: a short review. *Microporous and Mesoporous* Materials, 144, pp. 15–18.

Mossman, D. J., Lafaye, F. G., Dutkiewicz, A., and Bruning, R. (2008). Carbonaceous Substances In Oklo Reactors. Analogue For Permanent Deep Geological Disposal Of Anthropogenic Nuclear Waste. *Reviews in Engineering Geology*, XIX, pp. 1–13.

Papalan, R. T., Turner, D. R., Bertetti, F. P., and Prikryl, J. D. (1998). Uranium^{VI} Sorption Onto Selected Mineral Surfaces. In Jenne, E. A. (ed.), *Adsorption of Metals by Geomedia*, Chapter 3. Academic Press, San Diego, California, pp. 99–130.

Peterman, Z. E. and Cloke, P. L. (2002). Geochemistry of Rock Units at The Potential Repository Level. Yucca Mountain, Nevada. *Applied Geochemistry*, 17, pp. 683–698.

Posiva Oy. (2012). *Safety Case for the Disposal of Spent Nuclear Fuel at Olkiluoto — Design Basics 2012*. Eurajoki, Finland (Report number FIN-27160).

Posiva Oy. (2017). *ONKALO —The Underground Rock Characterisation Facility at Olkiluoto*, Eurajoki, Finland (Report number ENG 290306).

Preuss, J., Eilers, G., Mauke, R., Muller-Hoeppe, N., Engelhardt, H. J., Kreienmeyer, M., Lerch, C., and Schrimpf, C. (2002). Post Closure Safety of The Morsleben Repository. In *Waste Management Conference*, Tucson, Arizona, Feb. 24–28, 2002.

Ramana, M. V. (2013). Shifting Strategies and Precarious Progress: Nuclear Waste Management In Canada. *Energy Policy*, 61, pp. 196–206.

Raven, K. G., Sterling, S. N., Jackson, R.E., Avis, J. D., and Clark, I. D. (2010). Geoscientic Site Characterization of The Proposed Deep Geologic Repository, Tiverton, Ontario. *Canadian Society of Petroleum Geologists, GeoCanada 2010. Working with the Earth, Calagary, May 10-13, 2010.*

Rechard, R. P. (2014). Results From Past Performance Assessment For The Yucca Mountain Disposal System For Spent Nuclear Fuel And High-Level Radioactive Wastes. *Reliability Engineering and System Safety*, 122, pp. 207–222.

Rechard, R. P. and Stockman, C. T. (2014). Waste Degradation And Mobilization In Performance Assessments For The Yucca Mountain Disposal System For Spent Nuclear Fuel And High-Level Radioactive Waste. *Reliability Engineering and System Safety*, 122, pp. 165–188.

Rechard, R. P., Liu, H. H., Tsang, Y. W., and Finsterle, S. (2014). Characterization of Natural Barrier of Yucca Mountain Disposal System For Spent Nuclear Fuel And High-Level Radioactive Waste. *Reliability Engineering and System Safety*, 122, pp. 32–52.

Rempe, N. T. (2007). Permanent Underground Repositories for Radioactive Waste. *Progress in Nuclear Energy* 49, pp. 365–374.

Riplens, M. and Biurrun, E. (2002). Decommissioning and Closure Of The Morsleben Deep Geological Repository. The final step. In *Waste Management Conference*, Tucson, Arizona, February 24–28, 2002.

Rosborg, B. and Werme, L. (2008). The Swedish Nuclear Waste Program And The Long-Term Corrosion Behavior Of Copper. *Journal of Nuclear Materials*, 379, pp. 142–153.

Roxburgh, I. S. (1987). *Geology of High-Level Nuclear Waste Disposal: An Introduction*. Chapman and Hall, London, UK.

Rumynin, V. G., Sindalovskiy, L. N., Konosavsky, P. K., Mironova, A. V., Zakharova, E. V., Kaimin, E. P., Pankina, E. B., and Zubkov, A. A.

(2005). Review of the Studies Of Radionuclide Adsorption/Desorption With Application To Radioactive Waste Disposal Sites In The Russian Federation. *Developments in Water Science*, 52, pp. 271–311.

Rybalchenko, A. I., Pimenov, M. K., Kurochkin, V. M., Kamnev, E. N., Korotkevich, V. M., Zubkov, A. A., and Khafizov, R. R. (2005). Deep Injection Disposal of Liquid Radioactive Waste in Russia, 1963–2002: Results And Consequences. *Developments in Water Science*, 52, pp. 13–19.

Schwartz, M. O. (2009). Modelling Groundwater Contamination Above the Asse 2 Medium-Level Nuclear Waste Repository, Germany. *Environmental Earth Sciences*, 59, pp. 277–287.

Skogsberg, M. and Ingvarsson, R. (2006). Operational Experience from SFR–Final Repository for Low- and Intermediate-Level Waste in Sweden. *European Nuclear Society*, Topseal 2006, Olkiluoto, Finland, September 17–20, 2006.

Smellie, J. (2009). Analogue Evidence from Uranium Orebodies. Nuclear Decommissioning Authority. Cumbria, UK. Available at: http://www.nda.gov.uk/documents/ [Accessed 9 February 2020].

Stille, H. and Fredriksson, A. (1988). Measurements, Calculations, and Stability Prognoses at the SFR Undersea Repository for Low- and Medium-Level Nuclear Waste. *Tunneling and Underground Space Technology*, 3, pp. 277–282.

Swift, P. N. and Corbet, T. F. (2000). The Geologic and Hydrogeological Setting of the Waste Isolation Pilot Plant. *Reliability Engineering and System Safety*, 69, pp. 47–58.

Swift, P. N., Hansen, C. W., Helton, J. C., Howard, R. L., Knowles, M. K., MacKinnon, R. J., McNeish, J. A., and Sevougian, S. D. (2014). Summary Discussion of the 2008 Performance Assessment for the Proposed High-Level Radioactive Waste Repository at Yucca Mountain, Nevada. Reliability Engineering and System Safety, 122, pp. 449–456.

Szakálos, P. and Seetharaman, S. (2012). *Corrosion of Copper Canister*. Swedish Radiation Safety Authority (Technical Note 4).

US DOE. (2020). *The U.S. Department of Energy Waste Isolation Pilot Plant*. Available at: http://www.wipp.energy.gov/ [Accessed 9 February 2020].

US Senate Committee on Environmental and Public Works. (2006). *Yucca Mountain: The Most Studied Real Estate on the Planet.* Report to the Chairman Senator James M. Inhofe U.S. Senate Committee on Environment and Public Works Majority Staff.

Vaniman, D. T., Chipera, S. J., Bish, D. L., Carey, J. W., and Levy, S. S. (2001). Quantification of Unsaturated-Zone Alteration and Cation Exchange in Zeolitized Tuffs at Yucca Mountain, Nevada, USA. *Geochimica et Cosmochimica Acta*, 65, pp. 3409–3433.

Vinsot, A., Mettler, S., and Wechner, S. (2008). In Situ Characterization of the Callovo-Oxfordian Pore Water Composition. *Physics and Chemistry of the Earth*, 33, pp. 575–586.

WNR. (2019). *Country Profiles. World Nuclear Association.* Available at: http://www.world-nuclear.org/ [Accessed 9 February 2020].

Chapter 7

Managing Decommissioning Wastes

"At first I saw absolutely nothing. When I was able to open my eyes, I stood still, far more stupefied than astonished."

— Henry Lawson in *Journey to the Center of the Earth* by Jules Verne

7.1 Introduction

All types of nuclear-energy-related facilities have a finite lifespan. It is essential for the future of nuclear energy that these facilities be successfully and safely decontaminated and decommissioned when they reach the end of their operating life. Unlike the decommissioning and demolition of office buildings, bridges, or closed factories, radioactive materials are present at nuclear facilities. The occurrence of radiation sources increases the complexity of the planning and execution of the process. Decommissioning was defined by ITRC (2008) as "actions taken at the end of the life of a nuclear facility to retire it from service." Decontamination is "the removal or reduction of radioactive and/or other hazardous contamination." Decommissioning and decontamination are often abbreviated as "D&D." The "D" term, however, is sometimes used to mean deactivation, dismantlement, or disposition.

Internationally, more than 115 commercial power plants, 48 experimental reactors, 250+ research reactors and several fuel-cycle facilities have been retired from service and are at some stage of D&D

(WNR, 2018). There are several reasons for the removal of a nuclear facility from active service (in part, from Rahman, 2008):

1. *End-of-operating-life* as defined by the license granted by the regulatory agencies. If the license is not extended, the facility will need to be taken out of service.
2. *Uneconomical operation* because of increasing operating costs and/or decreasing revenues earned by the facility. The competition of other sources of energy can also be a major factor.
3. *Technical obsolescence.* The facility may become too old to justify major upgrades. Also, in the case of experimental reactors, when all the planned experiments are completed, and the lessons have been learned, the reactor may be shut down because it is no longer needed for future research.
4. *Change in governmental priorities or policies.* Nuclear facilities such as experimental reactors may be shut down because the government has lost interest and/or withdrawn funding for future operations. Note that some governments such as Germany have chosen to abandon nuclear energy which forces relevant facilities to close.
5. *Safety concerns.* A facility may be shut down permanently following a major incident. The facility may have suffered significant damages and/or the credibility of the installation (or that of nuclear energy in general) have been damaged in the eyes of the public.
6. *Obsolescence by redundancy.* Nuclear facilities may be shut down because they no longer meet a national priority. The end of the Cold War reduced the need for the Nuclear Weapons Complex in the US. This national network included more than 100 sites that mined, milled, refined, and enriched uranium, and produced plutonium, in addition to various facilities used to fabricate and test nuclear weapons (see Chapter 9). The D&D of these facilities is on-going and will continue well into the 21st century (US DOE, 2018).

The International Atomic Energy Agency (IAEA) has defined three options for decommissioning which have been adopted internationally (Rahman, 2008; WNA, 2018). These options are as follows:

1. *Immediate Dismantling* (DECON in the US). Soon after the nuclear facility is closed, there is a sequence of events such as the permanent removal of the reactor fuel from the reactor vessel. Other major events include the removal of the reactor vessel, steam generators, piping systems, pumps and valves. The primary motivation for selecting this option is to terminate the license as soon as possible.
2. *Safe Enclosure* (SAFSTOR in the US). This option is often considered as "delayed dismantling." It is typically applied to power plants. The facility is closed, and the reactor fuel is removed. Then, the facility is monitored for a period of time to allow the radioactive sources to decay to less problematic levels. Taking advantage of radioactive decay will reduce the risk of occupational exposure during D&D and can also reduce the costs of the disposal of the wastes. For example, some of the activation products in the reactor components have relatively short half-lives. Following a decay period, the D&D process will continue. The licensee has the option of a hybrid approach by using both approaches (immediate and delayed) to various areas during a D&D project.
3. *Entombment* (ENTOMB in the US). In this approach, radioactive components are left on-site, but are permanently encased in concrete. The facility is then monitored until the radioactivity has decayed to acceptable levels. This option has not been used frequently. However, eight of the nine reactors at the Hanford Nuclear Reservation have been "cocooned." These reactors were used during the Cold War solely to produce plutonium. Each reactor building was demolished, leaving only the core of the reactor. Each core is encased in concrete and steel. These cocoons will remain in place for about 75 years, after which the reactors

will be removed (ITRC, 2008). The Hallam Nuclear Generating Station in Nebraska was decommissioned in about 1969. The above-ground portion of the facility was demolished, but some radioactive materials were entombed below ground. Each of three storage cells is steel lined and covered with concrete. About 11,100 TBq (300,000 Ci) of radioactive material were initially entombed, and it was estimated that the site can be released for unrestricted use in 2070 (US DOE, 2017). Perhaps the most widely known example of entombment is at the Chernobyl Nuclear Power Plant in the Ukraine. Following an explosion, the remains of Reactor 4 were enclosed by a large concrete sarcophagus (Bradley, 1997). In 2017, the New Safe Confinement structure — composed of steel and concrete — was slid over the sarcophagus (from NucNet, 2018. Available at: https://www.nucnet.org/ [Accessed 9 February 2020].

The scope of this chapter is primarily about the chemical and physical methods available for decontaminating radiologically contaminated materials in the context of waste management. Regulatory perspectives as they relate to D&D will not be discussed. The reader can find information about regulations, guidelines, demolition planning and execution, project management and funding, and historical experiences and lessons learned as they relate to D&D projects in several publications such as IAEA (1999), Saling and Fentiman (2001), Taboas *et al.*, 2004; ITRC (2008), Rahman (2008), and NEA (2011). Three case examples are provided in this chapter to illustrate the sequence of events during the D&D of a nuclear facility. These are the Nuclear Research Reactor Facility of the University of Illinois, the Connecticut Yankee Nuclear Power Plant (US), and the Oskarshamn Nuclear Power Plant in Sweden.

7.2 Decontamination

Decontamination can be defined as the removal of radiological contaminants from the surfaces of facilities and equipment by a variety of chemical and physical techniques (US EPA, 2006). Decontamination in any form does not eliminate radioactivity, but simply transfers it from one location to another where it can be removed and isolated to protect human health and the environment. The selection of the method(s) to be used to decontaminate surfaces will depend on the surfaces (smooth or porous) and the physicochemical forms of the contaminants, in addition to the goal of the process such as disposal, recycling, or reuse. In D&D jargon, the contaminants are often classified as loose and fixed (Kinnumen, 2008; Rahman, 2008). This classification approach can be refined as follows:

1. Loose debris such as airborne particles, dried residues, chemical precipitates, particles trapped in pore spaces or surface fractures, and any weakly adhered masses.
2. Fixed surface coatings or adherent layers such as chemical alteration products resulting from oxidation or corrosion that may contain radioactive sources.
3. Fixed contaminants also include subsurface neutron activation products that formed during the operational period of the facility, and other contaminants that may have diffused into the matrix (Fig. 7.1).

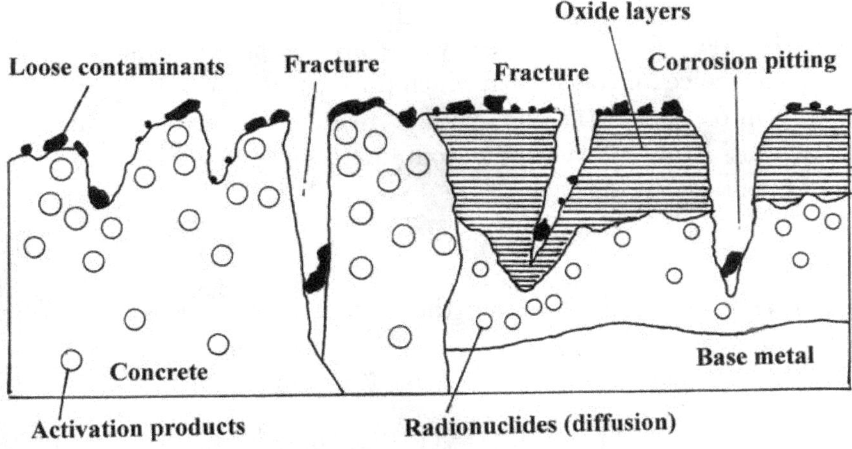

Figure 7.1. A hypothetical cross-section of surface contamination. Concrete is on the left, and metal is on the right. The details were not drawn to scale.

There are several potential benefits to decontaminating surfaces (Saling and Fentiman, 2001; US EPA, 2006) as follows:

1. Reduce the occupational exposure to radiation during decommissioning via inhalation, dermal contact and direct exposure.
2. Reduce the amount of radioactive waste that requires expensive disposal.
3. Enable the reuse of facilities and equipment.
4. May also remove non-radioactive contaminants.

Chemical decontamination is the application of various reagents dissolved in water, foams, and gels to extract contaminants from surfaces. Chemical decontamination is often accomplished by circulating solutions containing the reagents in closed loops. Segmented parts can be decontaminated by immersion (Kinnunen, 2008). Physical decontamination is the application of physical energy such as water flushing and spraying, vacuuming, grinding, scabbling, and shaving to remove contaminated materials from surfaces. Chemical decontamination is generally most effective on metallic and non-porous surfaces (Rahman, 2008). For effective chemical decontamination, the composition of the contaminated surfaces must be known (Kinnunen, 2008). Physical decontamination can be effective on almost any type of surface (US EPA, 2006). Physical approaches may be the only practical choice for concrete in which the contaminants have migrated deeply into the matrix and are not readily available for chemical extractants. The efficacy of decontamination can be described by the Decontamination Factor (DF), viz.,

$$\text{DF} = r_b/r_a \tag{7.1}$$

where r_b is the radiation from the surface before decontamination and r_a is the radiation from the same surface after treatment.

The reduction in radiation (Table 7.1) can be calculated as

$$\text{Reduction (\%)} = (1 - (1/\text{DF})) \times 100 \tag{7.2}$$

The selection of the decontamination technique often depends on the anticipated DF value needed, the compatibility of the material with the method, and the volume of secondary waste that will be

Table 7.1. The relationship between DF and the reduction in radiation.

DF	% Reduction
2	50
5	80
10	90
50	98
100	99

created by the application of the method (Kinnunen, 2008). The anticipated chemical composition and the methods available to treat the secondary wastes can also be critical to the selection process.

Both chemical and physical decontamination approaches have advantages and disadvantages (US EPA, 2006):

Advantages of chemical techniques

1. Most can be relatively simple, rapid, and inexpensive. The solutions can be applied as sprays, flushing discharges, and by immersion.
2. Chemical techniques can yield DF values that are greater than 10,000.
3. They can be used to treat surfaces that are physically difficult to reach with equipment typically used for physical decontamination.

Disadvantages of chemical techniques

1. Chemical decontamination can generate a liquid waste stream that must be captured and treated. Depending on the type of solution used, the treatment can include pH neutralization, ion exchange, chemical precipitation, and evaporation. These treatment approaches will yield secondary wastes that will contain the radionuclides or other contaminants. The secondary wastes include spent ion exchange resins, chemical precipitates, filtrates and spent filter membranes.

2. There is a risk of accidental releases of a liquid-treatment solution from spills or malfunctioning hose connections.
3. Heat may need to be applied to the extracting solution to accelerate the rate of the chemical reactions.
4. As mentioned earlier, chemical approaches may not be effective on porous surfaces.

Advantages of physical techniques

1. They can work on almost any type of surface.
2. They may be the only practical choice for porous concrete.
3. Physical decontamination may yield DF values that are greater than those possible with chemical approaches.
4. Waste management tends to be simpler because no secondary wastes are created that need to be managed.

Disadvantages of physical techniques

1. They are generally destructive to the surface being treated.
2. Airborne emissions of abraded particles (dust) must be controlled.
3. Inaccessible surfaces or complicated areas can be difficult for workers to reach.
4. They may require workers to operate tools closer to the contaminated areas which could result in larger doses of radiation.
5. The volume of waste may be larger if relative deep-surface removal is required, plus the volume of the abrasion media mixed with the decontaminated material.

7.3 Chemical Decontamination

7.3.1 *Strong Inorganic or Mineral Acids*

Inorganic acids are effective solvents for dissolving corrosion products and have been used as decontamination agents for years (Chen *et al.*, 1997). They can be used in concentrated forms or as dilutions with water. Nitric acid (HNO_3), hydrochloric acid (HCl), sulfuric acid (H_2SO_4), hydrofluoric acid (HF), phosphoric acid (H_3PO_4), fluoroboric acid (HBF_4), and sulfamic acid (H_3NSO_3) have all been used as decontamination agents separately or in combinations with each other and with other reagents.

Each acid will dissociate in water providing hydrogen ions (H^+) that react with surface metal oxides on stainless steel, carbon steel, aluminum and copper alloys. As these oxides dissolve, radionuclides that were sequestered by them can be mobilized and extracted. For example, the reaction of hydrogen ions with an iron oxide (magnetite) containing cesium:

$$8H^+ + (Fe_3O_4) - Cs \rightarrow 2Fe^{3+} + Fe^{2+} + Cs^+ + 4H_2O \qquad (7.3)$$

Nitric acid is a strong oxidant and is used to dissolve metallic oxide films on stainless steel and Inconel (a corrosion-resistant nickel–chromium–iron alloy) (IAEA, 1999), but may be too corrosive for carbon steel (US EPA, 2006). Nitric acid has been used to dissolve fission products and U and Pu oxides, on stainless steel during the decommissioning of the Eurochemic reprocessing plant in Belgium (Chen *et al.*, 1997). A 10% solution applied at about 75°C is typically used (US EPA, 2006).

Hydrochloric acid is used to remove scale (often called limescale) deposits of calcium and magnesium carbonates on large pieces of equipment in non-nuclear applications. Because of the corrosive nature of chloride ions, however, hydrochloric acid has not been widely used in the nuclear industry (Chen *et al.*, 1997).

Sulfuric acid is also used to remove iron-oxide deposits on stainless steel and carbon steel (Chen *et al.*, 1997). A limitation in its use, however, is that it cannot be effectively applied to deposits containing Ca, Mg, Ba, and Sr. The dissolution of such deposits can result in the precipitation of relatively insoluble sulfate compounds, viz.,

$$H_2SO_4 + CaCO_3 \rightarrow CaSO_4 \downarrow + CO_2 + H_2O \qquad (7.4)$$

Another limitation may be that relatively small DF values have been reported from the use of sulfuric acid (Chen *et al.*, 1997).

Both hydrofluoric and fluoroboric acid are capable of dissolving surface oxides and other impurities on stainless steel and carbon steel (IAEA, 1999; US EPA, 2006). Both acids are effective in dissolving silica-containing deposits. For example, the reaction of hydrofluoric acid with silicon dioxide:

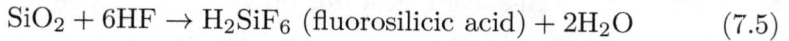

$$SiO_2 + 6HF \rightarrow H_2SiF_6 \text{ (fluorosilicic acid)} + 2H_2O \quad (7.5)$$

Phosphoric acid is a weak but corrosive acid. Dilute solutions (8–10% at 60°C–85°C) have been used to remove uranium oxides, fission products, and activation products from carbon steel surfaces and Zircaloy fuel surfaces (Chen *et al.*, 1997; IAEA, 1999). DF values of 10 or greater have been achieved. Phosphoric acid may be used in lieu of hydrochloric acid to avoid the chloride ions.

Sulfamic acid is a strong acid that is used as a cleaning agent for metals and ceramics. It has been used to decontaminate carbon steel when applied as an 8% solution at 85°C. The limited experience available suggests that sulfamic acid may require longer contact times with contaminated surfaces and repeated applications, when compared with other mineral acids (Chen *et al.*, 1997).

7.3.2 *Organic Acids and Chelation Agents*

Simple organic acids and chelation agents (also called complexing agents, chelators, and chelates) are used separately or in combination with each other and with mineral acids for the decontamination of surfaces. For example, citric acid ($C_6H_8O_7$) is both an organic acid and a chelation agent. The acidity results from the deprotonation (the loss of hydrogen) from the single-bonded hydroxides (Fig. 7.2(a)), resulting in the formation of its conjugate base (Fig. 7.2(b)).

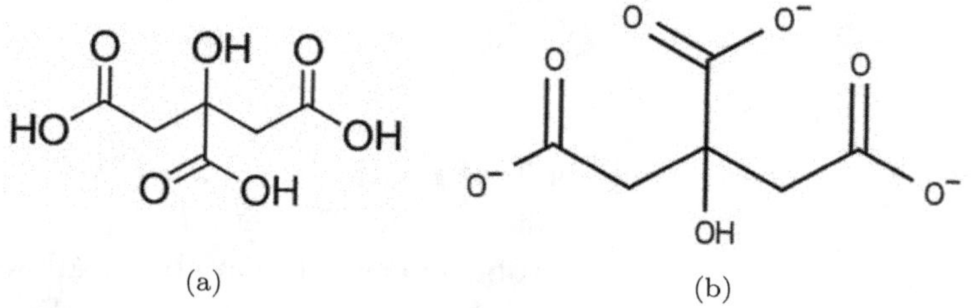

Figure 7.2. Citric acid (a) and its conjugate base, the citrate anion (b).

A chelation agent can form electrostatic bonds between the negative ions on the chelate, and positively charged ion (cations) such as Cs^+. Large chelation agents can form multiple bonds to a single cation, thus sequestering or shielding the cation from reacting with other anions in solution. Chelation agents promote the solubility of contaminants by preventing them from precipitating which facilitates their removal during decontamination.

Citric acid is used to dissolve iron oxides from stainless steel (IAEA, 1999; Rahman, 2008). Citric acid in combination with sulfuric acid was used to remove uranium and technetium from metal surfaces during the decommissioning of the Capenhurst Gaseous Diffusion Plant in the UK (Chen *et al.*, 1997). A mixture of citric and oxalic acid ($C_2H_4O_4$) has also been used to decontaminate stainless steel. Oxalic acid is also a chelation agent and has been used to complex niobium and fission products (IAEA, 1999).

One of the best-known chelation agents is ethylenediaminetetraacetic acid (EDTA) ($C_{10}H_{16}N_2O_8$). It is used in medicine, cosmetics, food preservation, industrial applications, and during the decontamination of surfaces. Its conjugate base, ethylenediaminetetraacetate ($EDTA^{4-}$), can bond with cations by electrostatic interactions with two amines and four carboxylates (Fig. 7.3).

EDTA and other chelation agents such as diethylenetriaminepentaacetic acid and hydroxyethylethylenediaminetriacetic acid are

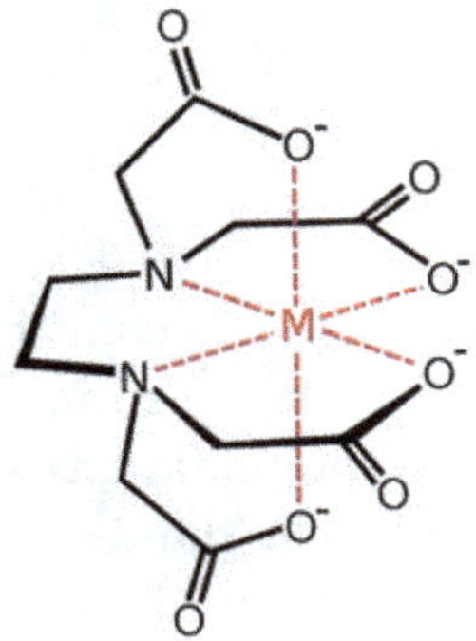

Figure 7.3. The ethylenediaminetetraacetate anion and a metal cation (M) forming a six-coordinate complex.

used during decontamination to prevent the redeposition of metals ions extracted on metal surfaces (Rahman, 2008). EDTA has proven to be effective in some situations when mixed with citric acid on stainless steel (Chen *et al.*, 1997).

7.3.3 *Oxidizing and Reduction (Redox) Agents*

Redox is a type of chemical reaction in which a reducing agent donates electrons, while the oxidizing agent accepts the electrons. There must always be an oxidant and a reductant for the reaction to occur. The redox state of an element (whether it is oxidized or reduced) can be important in decontamination. Some metals are more soluble in solution in reduced forms when compared with their oxidized states. Iron for example is more soluble as Fe^{2+} under reduced conditions, whereas Fe^{3+} dominates in oxidized solutions, and may precipitate as ferric hydroxide (Roy, 2005). In other situations, however, the more oxidized forms of a metal will have a greater positive charge and may be more reactive with chelation agents such as EDTA. It follows that the ability to alter or control the redox state of the extracting fluid can influence the efficacy of decontamination.

The redox state of an extraction scenario can be chemically reduced by adding a reducing agent which acts as a source of electrons and becomes oxidized by the reaction. Conversely, an oxidant becomes chemically reduced. Examples of reducing agents include oxalic acid ($C_2H_2O_4$) and formic acid (CH_2O_2). In the case of formic acid, the redox half-reaction can be written as

$$HCOOH \rightarrow CO_2 + 2H^+ + 2e^- \tag{7.6}$$

Oxidizing agents include household bleach which is a mixture of sodium hypochlorite (NaOCl), calcium hypochlorite ($Ca(ClO)_2$), and sodium hydroxide (NaOH). An important oxidant that is often used in chemical decontamination is the permanganate ion MnO_4^-. Because manganese is in the 7+ oxidation state, it readily accepts electrons. In acidic solutions, it is reduced to divalent manganese, viz.,

$$8H^+ + MnO_4^- + 5e^- \rightarrow Mn^{2+} + 4H_2O \tag{7.7}$$

In alkaline solutions, it is reduced to MnO_4^{2-}, and in pH-neutral solutions, the reduced form of manganese is MnO_2 which can precipitate as a solid.

Cerium as the cation Ce^{4+} has been used as an oxidant in decontamination when combined with other reagents. The half-reaction is simply

$$Ce^{4+} + e^- \rightleftarrows Ce^{3+} \tag{7.8}$$

Trivalent cerium becomes the reduced byproduct of the reaction. Cerium nitrate, dissolved in nitric acid, was used to decontaminate steel glove boxes at the former Rocky Flats Plant in Colorado. The glove boxes were contaminated with plutonium, americium, and uranium. The application of the cerium nitrate allowed the gloveboxes to be disposed as low-level radioactive wastes (GAO, 2006).

7.3.4 *Combined-Treatment Processes*

Several combined processes have been developed, tested, revised, and commercialized for decontamination. These processes can be combinations of strong mineral acids, organic acids, chelation agent, and redox agents that are applied in multiple steps. It is beyond the scope of this chapter to describe all these processes. Chen *et al.* (1997) and Kinnunen (2008) provided a comprehensive list of many combined-treatment processes. For this chapter, five processes were chosen for discussion that should illustrate the diversity in their approaches.

1. Alkaline Permanganate-Citric and Oxalic acids (AP-CITROX) process has been used to remove corrosion films from high-chromium stainless steel. For example, Varga *et al.* (2001) conducted a laboratory study on contaminated stainless-steel samples from the Paks Nuclear Power Plant in Hungary. They used a 3-step application of sodium hydroxide + potassium permanganate followed by citric acid + oxalic acid. The final treatment was hydrogen peroxide (H_2O_2) + ammonia (NH_3), all applied at 90°C.

The use of permanganate is also part of other processes such as APAC (alkaline permanganate–ammonium citrate), APACE (alkaline permanganate–ammonium citrate–EDTA), and APOX (alkaline permanganate–oxalic acid).

2. Chemical Oxidation Reduction Decontamination (CORD). The basic CORD approach was developed in Germany (Chen *et al.*, 1997), and has since evolved in the Cord Family of procedures. These variations have gained international acceptance. Various CORD-based procedures have been used in Belgium, Germany, Finland, Italy, Japan, Sweden, and the USA (IAEA, 1999; Kinnunen, 2008). One such variation is called HP/CORD D UV. It consists of:

   A. A pre-oxidation step using potassium permanganate to dissolve chromium-containing oxides.
   B. Oxalic acid to reduce the permanganate, and to dissolve contaminants.
   C. Decomposition of the decontamination agents by using ultraviolet light and hydrogen peroxide — all applied at 95°C (Kinnunen, 2008).

3. Decontamination for Decommissioning (DFD). This procedure was developed by the Electric Power Research Institute in 1996. It uses a combination of potassium permanganate, oxalic acid, and fluoroboric acid (Noynaert, 2012). The DFD Process was designed to decontaminate stainless steel and Alloy 600 components for unrestricted release (EPRI, 2004). It has been applied to the cooling systems of the now decommissioned Yankee Power Plant (Bushart and Wood, 2003). The DFD procedure has been revised and renamed the EPRI DFDX to include a treatment process for the secondary waste created by the DFD process. The new treatment process is a combination of cation and anion exchange resins coupled with an electrochemical reaction to convert metallic ions into neutral metallic particles. Pilot-scale experiments of the revised procedure were described in EPRI (2004).
4. Low Oxidation State Metal Ion (LOMI). This process was first developed in the UK, and there are various LOMI-based processes

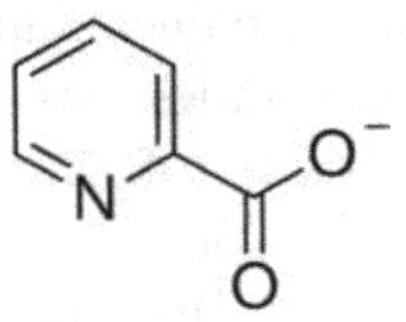

Figure 7.4. Picolinate, the conjugate base of picolinic acid.

have evolved (Chen *et al.*, 1997). Vanadium (II) is used as a reducing agent and picolinic acid ($C_6H_5NO_2$) is added as a chelation agent. The conjugate base of the latter is picolinate (Fig. 7.4). LOMI has been used to dissolve iron oxide films on metal surfaces (Noynaert, 2012). Applied as a heated solution (80°C–90°C), the vanadium reduces Fe^{3+} to the more soluble Fe^{2+}. The half-reaction of vanadium is

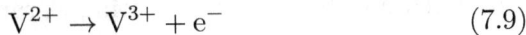

$$V^{2+} \rightarrow V^{3+} + e^- \tag{7.9}$$

The half-reaction for iron reduction is

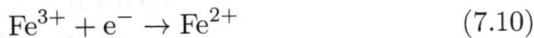

$$Fe^{3+} + e^- \rightarrow Fe^{2+} \tag{7.10}$$

When the two half-reactions are combined, we have

$$V^{2+} + Fe^{3+} \rightarrow V^{3+} + Fe^{2+} \tag{7.11}$$

The LOMI process has been successfully applied at the Pacific Northwest National Laboratory (Washington, US), the now decommissioned Winfrith Experimental Reactor in the UK, and the Monticello Nuclear Generating Plant in Minnesota (US) (Chen *et al.*, 1997). The typical LOMI process, however, may not be suitable for oxides in pressurized-water reactor systems (Kinnunen, 2008).

5. MEDOC (Metal Decontamination by Oxidation with Cerium) was developed in Belgium. It is a process that uses Ce^{4+} as an oxidant, and sulfuric acid as a solvent (Ponnet *et al.*, 2000). Ozone is used to continuously regenerate Ce^{4+}:

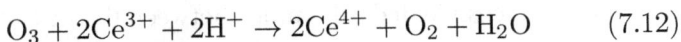

$$O_3 + 2Ce^{3+} + 2H^+ \rightarrow 2Ce^{4+} + O_2 + H_2O \tag{7.12}$$

MEDOC was used during the decommissioning of Belgium Reactor 3. When applied to stainless steel piping, tanks, and heat exchangers, the radioactivity of 75% of the material treated yielded less than 0.1 Bq/g for cobalt-60 (Ponnet *et al.*, 2003). The MEDOC process can be used in batches of cut pieces or in a closed loop for decontamination. DF values greater than 10,000 are possible (Noyaert, 2012).

7.3.5 *Foams, Gels, and Strippable Coatings*

A feature common to all foams, gels and strippable coatings is that they are applied to contaminated surfaces as carriers of decontamination agents, but not act as the agents themselves. They are used to remove loose-surface contamination and have been used in the nuclear industry for decades, and new formulations are being developed and tested for marketing (Draine, 2009).

Foams can be applied as a thin layer to surfaces, then rinsed of with high-pressure water. In addition to foaming agents, additional chemicals have included sodium hydroxide to create an alkaline foam, and sulfuric acid or nitric acid to generate an acidic foam. To the latter, cesium as $Ce(SO_4)_2$ has been added as an oxidant (reaction (7.8)) (Chen *et al.*, 1997). Foams were used in the decontamination of walls at the West Valley Demonstration Project (US).

Gels and strippable coatings are applied using a brush, trowel, roller, or power sprayer. Ideally, the application is allowed to dry, then peeled manually from the surface in pieces or larger detached films (Fig. 7.5) (IAEA, 1999). Early gel formulations consisted of a non-ionic detergent mixed with a carboxymethycellulose gelling agent that contained nitric acid, hydrofluoric acid, and oxalic acid as the decontamination agents (Chen *et al.*, 1997). The first commercial strippable coating was ALARA™ 1146 (US DOE, 2000). Because of the proprietary nature of the gel, detailed information about its chemical composition is not available. However, it contains potassium olelate and buytyl ricinoleate (both surfactants) (Draine, 2009). Test data demonstrated that the vinyl coating when applied to metal or concrete surfaces yielded DF values between 6 and 7 at the Savannah

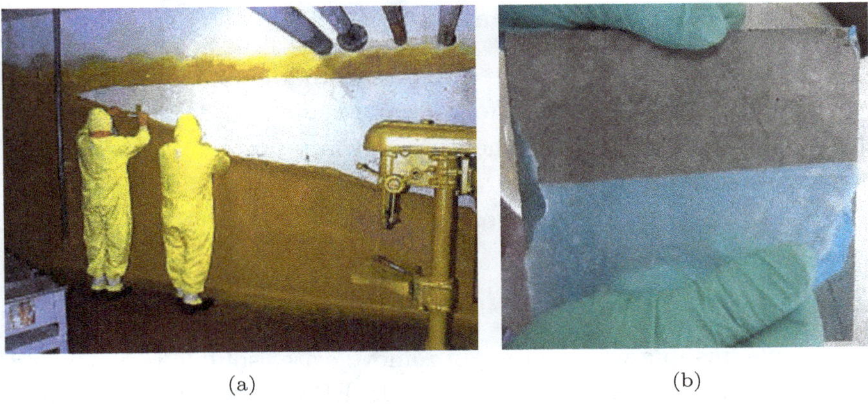

(a) (b)

Figure 7.5. ALARA 1167 (a) (US DOE, 2000) and DeconGel 1101 (b) (US EPA, 2011).

River Site (US). It has also been used during decontamination efforts at Los Alamos National Laboratory, the former Rocky Flats Plant (both US), Sellafield (UK), and Chernobyl (Ukraine) (IAEA, 1999).

Another example of a gel-based strippable coating is DeconGel™ 1101 (US EPA, 2011). This commercial product was first introduced in 2007. Because of the proprietary nature of the gel, detailed information about its chemical composition is not available. However, it contains a copolymer, ethanol, sodium hydroxide, and some type of chelation agent (Draine, 2009). Sutton *et al.* (2008) used DeconGel 1101 to remove plutonium from a contaminated glove box at Lawrence Livermore National Laboratory. A DF value of 130 was achieved on cast steel. DeconGel 1101 was used to remove technetium-99m, thallium-201, and iodine-131 from test tiles (vinyl and stainless steel) in a laboratory study (Draine, 2009). In a pilot-scale test, DeconGel 1101 and 1108 removed cesium-137 from concrete coupons. After two applications, DF values of 1.9 and 3.7, respectively, were achieved (US EPA, 2011). DeconGel 1101 was also tested successfully in removing cobalt-60 and cesium-137 from metal components collected from a research reactor in Romania to be decommissioned (Gurau and Deju, 2015).

7.4 Physical Decontamination

The various approaches and techniques available for physical decontamination can be grouped for discussion in this chapter by contaminant type of depth. The three groups are (1) loose-surface contaminants, (2) fixed contaminants at or close to the surface, and (3) deeper, subsurface fixed contaminants.

7.4.1 *Loose-Surface Contaminants*

1. Simple flushing with water can remove water-soluble residues and unbound particles from surfaces (US EPA, 2006). It is also used to as the preparatory step as a first wash (Noynaert, 2012). An increase in water pressure and volume will increase the efficacy of this approach. Common terms for this water-based application include hydroblasting, water jetting, and water-jet cutting at sufficiently strong pressures. Information about the extent of radiation reduction by only water-based methods is lacking (US EPA, 2006).
2. Dry vacuuming has been used successfully on radioactive-surface contamination. Dry vacuuming is typically used after the application of other physical approaches that create airborne byproducts. Dry vacuuming removes only loose particles and not fixed surface or subsurface contaminants (US EPA, 2006). If dry vacuuming is combined with water as steam, steam-vacuum cleaning become another option. This approach when used with a detergent can also be used to remove non-radiological contaminants and has been used on metal surfaces and concrete.

7.4.2 *Fixed-Surface Contaminants*

1. Common to all the techniques discussed in this section is that abrasive particles are pneumatically accelerated and directed at a surface. Grit or abrasive blasting can remove surface oxides and contaminants from cement, concrete, and steel (US EPA, 2006). The types of abrasives available are diverse, and include

sand, silicate minerals (staurolite, garnet, and olivine), hematite (an iron mineral), slags (derived from coal, copper, and nickel), crushed glass, and steel particles (rounded and angular shot).

2. Soft-media blast cleaning or so-called sponge blasting is another option. This approach uses particles of steel, garnet, aluminum oxides or Starblast^{TM} (staurolite) that are embedded within various types of plastic pellets. The resulting soft media do not bounce back from the treated surface readily, and can be collected for reuse (IAEA, 1999; US EPA, 2006). The efficacy of soft media was demonstrated as part of an effort to decontaminate materials at the Fernald Environmental Management Project (US).
3. Dry ice blasting is another option for surface decontamination (Fig. 7.6). Frozen carbon dioxide pellets are fired at a surface. The pellets remove the contaminants by three mechanisms: (1) the impact of the pellets, (2) the contrast in temperature between the pellets and the surface may disrupt the "bonds" holding the particles in place, and (3) the evaporation of the pellets into vapor may help pull the contaminants off the surface (Kinnunen, 2008). Although commercial equipment is available, there are few data available to document the efficacy of this technique (US EPA, 2006). It may not be effective for subsurface fixed contamination in concrete (NEA, 2011).

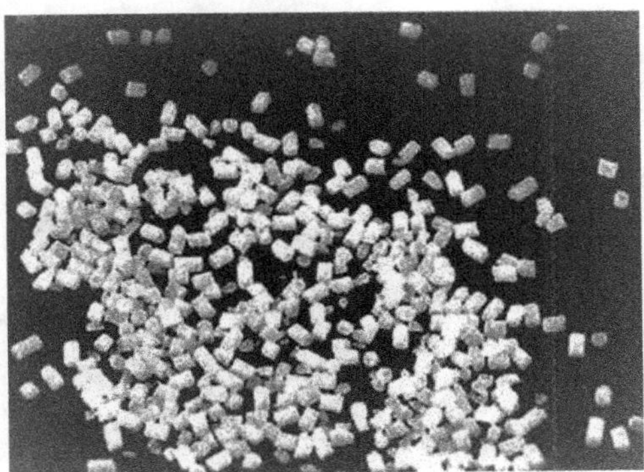

Figure 7.6. Dry ice pellets (US EPA, 2006).

7.4.3 *Fixed-Subsurface Contaminants*

1. The simplest technique is the use of a hand-held concrete grinder that has a diamond grinding wheel (Fig. 7.7). A grinder can remove concrete to a depth of about 1.5–3 mm (US EPA, 2006). It may be most applicable for small areas or for "hot-spot decontamination."
2. Shavers (Fig. 7.8) consist of a rotating drum with diamond-embedded, closely-spaced blades that cut into a concrete surface. The shaving unit can be self-propelled or operated manually. Concrete shavers can remove between 1 and 15 mm of concrete

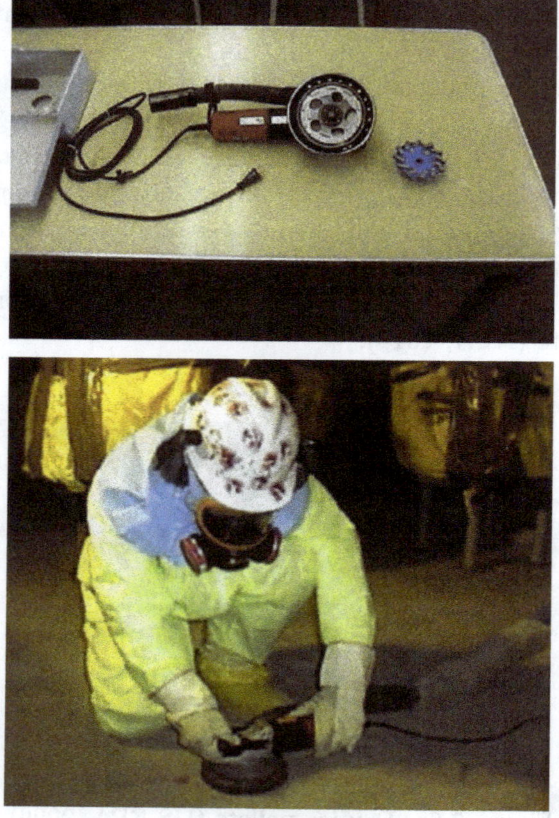

Figure 7.7. Hand-held grinders (US DOE 1998).

Figure 7.8. Underside of a floor shaver showing the diamond shaving blades (US EPA, 2006).

(IAEA, 1999; US EPA, 2006). Shavers are well suited for large, flat surfaces (NEA, 2011). A shaver was used during decontamination efforts at the Hanford Reactor C Building (US) (US DOE, 1998).

3. Scabblers (Fig. 7.9) are a group of pneumatically-driven devices that are used to remove layers of concrete by impacting the surface repeatedly with steel pistons or rods (IAEA, 1999; NEA, 2011). They can be used as hand-held devices or as larger machines that are manually or remotely operated. They can remove from 15 to 25 mm of concrete with multiple passes (Noynaert, 2012). Scabblers have been coupled with other approaches to make hybrid machines. For example, the En-vac Robotic Wall Scabbler uses grit blasting during scabbling. The Electro-Hydraulic Scabbler removes concrete under a thin layer of water using an electrical discharge to break the concrete. The Electro-Hydraulic Scabbler yielded a DF value of greater than 10 for uranium removal at the Fernald Environmental Management Project (US) (US EPA, 2006).

Figure 7.9. Floor scabblers (US EPA, 2006).

4. A concrete spaller is a hand-held device that removes relatively deep and large pieces of concrete. It drills holes into the surface, then a spaller bit is inserted into the hole. The bit expands and breaks the concrete. Spallers are intended for depths of 3 mm or deeper (IAEA, 1999; US EPA, 2006). Similarly, hydraulic or pneumatic hammering is regarded as a more rapid approach to remove contaminated concrete than scabbling and spalling (Noynaert, 2012). This approach can be accomplished with the familiar "jackhammer" or "pneumatic drill" (IAEA, 1999).

7.5 Three Case Studies of Nuclear Facility Decommissioning

7.5.1 *The Nuclear Research Reactor Facility*

The Nuclear Research Laboratory (NRL) was on the campus of the University of Illinois at Urbana-Champaign. It was in operation for about 30 years since the reactor first went critical in 1960. The reactor was a TRIGA Mark II training and research nuclear reactor (Fig. 7.10). The NRL was a steel frame, concrete building that was 13.7 by 24.4 m. The floor of the reactor room was a 15-cm concrete slab that was laid over undisturbed soil. There was a tunnel 4.6 m below grade that connected the reactor tank with the mechanical equipment room. The reactor was initially operated with a maximum power rating of 100 kW using 20% enriched uranium. By 1967,

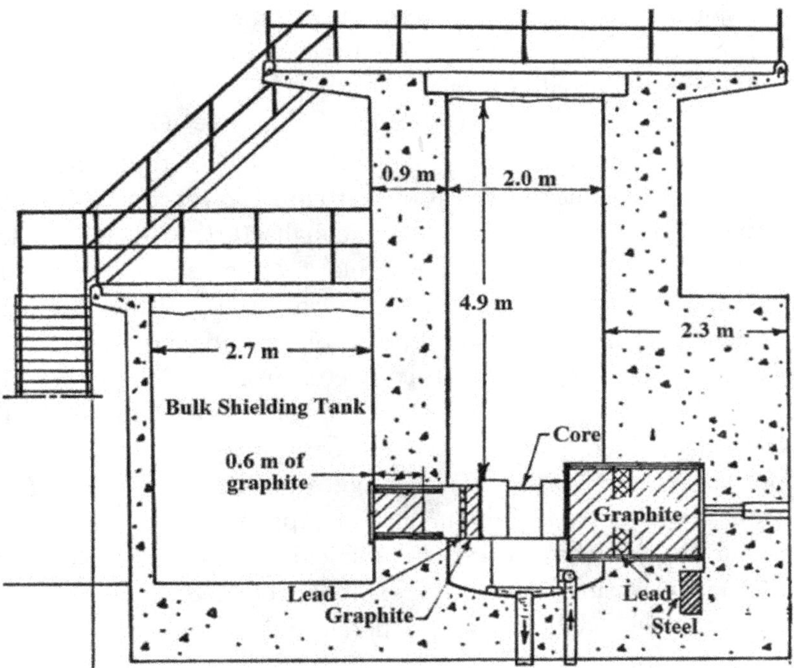

Figure 7.10. Cross-section of the bioshield of the TRIGA Mark II nuclear reactor, and sources of activated concrete, graphite, and steel (modified from an unpublished Site Characterization Report was issued by Scientech, LLC).

upgrades and license amendments allowed for the operating limit to be increased to 250 kW (Roy and Holm, 2013).

In 1998, The University administration decided to shut down the reactor permanently. In 1999, the reactor was placed in SAFSTOR. The Bulk Shielding Tank (Fig. 7.10) was used for storage of the spent nuclear fuel following the shutdown. The University planned to remove all radioactive materials from the NRL, demolish the building, and release the property for unrestricted use. In 2004, the spent fuel was removed and shipped to the US Department of Energy's Idaho National Laboratory. In 2005, a historical site assessment and a site characterization report were prepared to identify the various isotopes and estimate their activities at the NRL.

The major isotopes of concern were activation products, tritium, and fission products that might have been released during an experiment with uranium-235-coated tubes.

As expected, the concrete bioshield contained activation products. Thirty samples of the concrete revealed that there was asymmetrical "radius of activation" around the reactor orifice. The principle activation products were cobalt-60 (^{60}Co) most likely derived from trace amounts of stable ^{59}Co in the concrete, and europium-152 (^{152}Eu) and 154, likely from the activation of trace amounts of stable europium. Stable cobalt and europium can be found in almost any type of concrete. Iron-55 (^{55}Fe) was also present which was derived from the activation of stable iron. The concrete contained steel rebar, shadow shields, angle irons and supports, and magnetite (FeO_4) aggregate. The volume of the activated concrete was about 20.3 m^3, or about 71,700 kg. The total activity of the activated bioshield concrete was 3.11 GBq (84 mCi). Activation products were not detected in samples collected from the concrete floor, therefore there was no need for decontamination.

A thermal column for generating thermal neutrons penetrated the bioshield and core reflector assembly. All the reactor-grade graphite that was removed from the column was activated. The estimated total activity of the graphite removed was 0.26 GBq (7.03 mCi). The major activation products in the graphite were carbon-14 (^{14}C) and ^{152}Eu. Many of the reactor components were constructed of aluminum alloy. For example, the bioshield enclosed the aluminum reactor tank and the tank contained the reactor-core assembly. The mass of the activated portion of the tank was estimated as 212 kg. The major activation products of concern were ^{60}Co, ^{152}Eu, ^{55}Fe, and nickel-63 (^{63}Ni). A rotary specimen rack was used to irradiate materials and geologic samples for research applications. Although the specimen rack was composed of aluminum, it contained a chain drive with Stellite bearings. Stellite is a cobalt–chromium alloy designed for wear resistance. The otherwise stable ^{59}Co was activated into ^{60}Co. It was estimated that the total ^{60}Co activity of the rotary specimen rack was about 148 GBq (4 Ci), making it the most problematic component during the decommissioning of the NRL. Tritium was widespread; it was detected in the bioshield, the containment floor, subterranean walls and floors, and the subsurface soil.

After being in SAFSTOR for about 12 years, the decision was made in 2011 to proceed with the decontamination and decommissioning of the NRL. The NRL was prepared for decommissioning by first removing and packaging wastes in the building that were created from the operation or experiments conducted within the reactor. One of the large radioactive items was a glove box. Contamination by fission products was suspected, and composite smear samples were taken to assess the extent of removable contamination. The sum of fractions derived from plutonium-238 and 239, and nickel-63 was about 83% allowed for Class A low-level radioactive waste, and there was no need to decontaminate the glove box. After the glove box was removed, the process of removing the interference (items such as piping, ductwork that may hinder major decontamination or demolition activities) from the reactor tank was begun. The reactor components removed were the control rod drives, guide tubes, the reactor bridge, fuel racks, and cooling lines.

The rotary specimen rack (the "Lazy Susan") was then then removed. Because of the dose given off by the rack (0.6 Sv/h or 60 rem/h at the surface), there was the need to hoist the rack up from reactor tank using the overhead crane, then lower it into a shipping container as quickly as possible without delays to minimize occupational exposure. When the rotary specimen rack was lifted from the reactor tank (Fig. 7.11), and lowered into the shipping container, the process was accomplished in 86 seconds, which included a brief pause to measure dose while suspended above the bioshield (1.8 mSv/h or 180 mrem/h at a distance of 3 m).

Following the removal of the metal interference from the reactor tank, preparations to remove the cantilever and the upper portion of the bioshield began. However, before making the initial cuts in the cantilever, the top surface of the bioshield had to be decontaminated. Contamination levels of about 10,000 disintegrations per minute beta/gamma were reduced to non-detectable levels by scrubbing

Figure 7.11. Lifting the rotary specimen rack from the reactor tank (photograph taken by the author).

the surface with a solution containing sulfuric acid and proprietary surfactants.

Using air-cooled diamond core drills, a water-cooled track saw and diamond wire saws, the cantilever was cut into pieces. The upper bioshield and the Bulk Shielding Tank were then cut into blocks and removed (Fig. 7.12). The concrete pieces were disposed as non-radiological waste. A remote-controlled Brokk 330 with a demolition hammer attachment was used to rubblize the activated concrete within the bioshield. As the excavation progressed, activated metal components that were imbedded in the concrete such as the shadow shields were also removed. Approximately 43,137 kg of concrete rubble and rebar were placed in 15 Super Sacs (Fig. 7.12) and shipped to Energy*Solutions* Clive Facility in Utah (see Chapter 4) as a Class A waste.

About 227 m^3 of concrete waste were generated from the floor of the NRL and the bioshield demolition. Following the removal

(a) (b)

Figure 7.12. The upper bioshield being cut into blocks with a water-cooled concrete saw (a), and the Brokk 330 with Super Sacs (b). Photographs taken by the author.

of scrap metal and debris, the remaining bioshield was reduced to rubble, and the debris was removed, leaving an excavation that was about 3 m below grade. A Final Status Survey (FSS) documented that the NRL site compiled with the radiological release criteria that were approved by the US NRC for the license termination. As part of the survey, soil samples were collected on-site. Of the eight radionuclides chosen for the FSS (carbon-14, cesium-137, cobalt-60, europium-152 and 154, iron-55, nickel-63, and tritium), only ^{137}Cs was present in concentrations that were greater than the Minimum Detectable Concentration (1.85 mBq/g or 0.05 pCi/g). Although the ^{137}Cs could have been derived from the NRL, it was more likely from fallout during the above-ground testing of nuclear weapons from 1945 to 1962. The exaction was filled and revegetated. In 2015, the site was recognized by as a National Historic Landmark by the American Nuclear Society.

7.5.2 *The Connecticut Yankee Nuclear Power Plant*

The Connecticut Yankee Plant was in Haddam Neck, Connecticut (from EPRI, 2006).[1] It was first engaged for commercial operation in 1968 (Fig. 7.13) and was operated at 619 megawatts. In 1996,

[1]The plant was sometimes called the Haddam Neck Power Plant.

Figure 7.13. The Connecticut Yankee Nuclear Power Plant (EPRI, 2006). Used with permission of the Electric Power Research Institute.

Connecticut Yankee was permanently shut down because it was no longer economical to operate. The DECON option was chosen for the decommissioning of the plant. All the fuel assemblies were removed from the reactor vessel and placed in the spent fuel pool.

A CORD UV-based process was used for chemical decontamination. It was applied to the reactor-coolant system and part of the steam generator tubing. The anticipated benefits for chemical decontamination were to reduce occupational exposure during decontamination and a reduction in waste-disposal costs. It was estimated that the overall DF resulting from the CORD UV process was 15.9, and that a total of 4.77 TBq (129 Ci) of cobalt-60 was removed. It was also concluded that of the surfaces treated, about 90% of the radioactive dose was reduced. Care had to be taken to change the cation exchange resins frequently to prevent them from becoming GTCC wastes (see Chapter 4). The presence of transuranic radionuclides in oxide films resulted in the demand for the exchange resin used to treat the spent-treatment liquids.

The spent nuclear fuel was stored in the fuel pool. An effort was made to keep each batch of fuel in the pool for 5 years to allow the decay heat to decrease prior to the next stage. The spent fuel was then removed from the pool and placed into 40 NAC International, Multi-Purpose Canisters. Above ground storage and transport casks for spent nuclear fuel are discussed in Chapter 5. The canisters were placed in an Independent Spent Fuel Storage Installation (ISFSI) (Fig. 7.14). Such an installation is essentially is a thick concrete slab that is intended for the interim storage of spent fuel. The installation is constantly guarded and monitored. In addition to the 40 fuel casks, 3 casks contain GTCC wastes that were created during the dismantling operations. It is envisioned that the casks will be transported to a permanent geological repository when it becomes available (see Chapter 6).

The dismantling of the steam generator, reactor coolant pump, and the internal components in the reactor vessel was accomplished

Figure 7.14. The Independent Spent Fuel Storage Installation at Connecticut Yankee (EPRI, 2006). Used with permission of the Electric Power Research Institute.

by segmentation. The operation was conducted using a water jet with an abrasive. Garnet sand was used as the abrasive. The steam generator was cut into two segments. One segment was transported by rail to the Energy*Solutions* Clive Facility in Utah. The other segment was transported by both barge and rail to the Barnwell Waste Management Facility in South Carolina (see Chapter 4). The reactor vessel was transported by barge to Barnwell.

The containment building at Connecticut Yankee consisted of a reinforced concrete exterior with an inside liner made of carbon steel. The dome-topped building was 51.8 m (170 feet) tall, and 42.7 m (140 feet) in diameter. The demolition of the building was accomplished in steps. In each step, the bottom portion of the building was removed using excavators equipped with hoe rams to break up the concrete. After the removal of the base, the building was allowed to drop in place for the next step (Fig. 7.15). The concrete was rubbilized and removed. This sequence was repeated

Figure 7.15. Demolition of the containment building at Connecticut Yankee (EPRI, 2006). Used with permission of the Electric Power Research Institute.

Table 7.2. Partial Site Release Limits used for soil remediation at Connecticut Yankee (EPRI, 2006).

	Limit	
Radionuclide	**pCi (g)**	**Bq (g)**
Cesium-137	4	0.15
Cobalt-60	6	0.22
Strontium-90	23	0.85
Tritium	228	8.4

until the dome was lowered to the ground surface. The dome was then demolished in place.

The soils at the site were contaminated with radiation sources. A total of 33,000 m^3 (1,166,000 $feet^3$) was excavated and disposed off-site. The criteria used for soil excavations were release limits that were agreed by both the US EPA and the US NRC (Table 7.2). The derivation of these release limits was a complicated and iterative process. As an input to that process was the application of RESRAD, a dose modeling code that was developed by Argonne National Laboratory (US). The resident farmer scenario was chosen for the analysis. In this scenario, a farmer and his family grow their food on the post-remediation site and obtain their drinking and irrigation water from an on-site well. The reader will gain a better understanding of this type of dose modeling by solving Problem 12 in the Review Questions. This exercise will take the reader to a US EPA web-based model that calculates Preliminary Remediation Goals for site restoration.

The D&D of the Connecticut Yankee Nuclear Power Plant was completed in 2007. A post-remediation, groundwater monitoring program was completed in 2015. The final disposition of the former plant site remains uncertain. Some of the adjacent property has been transferred to the US Fish and Wildlife Service. The ISFSI will remain in service until a long-term option for waste management becomes available.

7.5.3 *The Oskarshamn Nuclear Power Plant, Sweden*

Swedish experience in decommissioning nuclear facilities is limited (SKB, 2005). The Oskarshamn Nuclear Power Plant (Fig. 7.16) was chosen for the chapter because it will serve as an example of future D&D operations in the 21st century. The plant consists of three units: Oskarshamn-1 (O-1), Oskarshamn-2 (O-2), and the adjacent Oskarshamn-3 (O-3). O-1 consists of a 472-megawatt boiling-water reactor, and O-2 is a 638-megawatt boiling-water reactor. O-1 and O-2 are now scheduled for decommissioning. O-2 was shut down in 2013 after 39 years of service. O-1 was terminated in 2017 after 45 years of power generation (WNA, 2018). Both units were shut

Figure 7.16. The Oskarshamn Nuclear Power Plant. From left to right, units O-1, O-2, and O-3 (Courtesy of SKB, Photo: Curt-Robert Lindqvist).

down because of economic reasons. O-3 will continue in service until about 2040.

The decision to terminate O-1 and O-2 was made years before the planned decommissioning dates (Rannemalm *et al.*, 2016). As a consequence of this early closure, the D&D planning process is currently in a state of flux. No decision has been made whether Immediate Dismantling or Delayed Dismantling will be implemented (Bergh *et al.*, 2014). Based heavily on international D&D experiences and the pioneering study of Larsson *et al.* (2013), numerous planning studies are being conducted by the Swedish Nuclear Fuel and Waste Management Company, Westinghouse Electric Sweden AB, and OKB AB. There are plans, for example, to combine O-1 and O-2 into a single D&D project (Rannemalm *et al.*, 2016). These plans will require approval by the Swedish Land and Environmental Court which will also include an Environmental Impact Assessment. The planning and approval process will likely require years before completion. At present, however, an anticipated sequence of events has been proposed (Bergh *et al.*, 2014; Rannemalm *et al.*, 2016):

1. The spent nuclear fuel will be removed from the plants and transported to the Central Holding Storage for Spent Nuclear Fuel (see Chapter 6). The units will become dormant but maintained as the planning and approval process continues.
2. Removal of the internal parts of the reactor vessel and part of the turbine plant. Decontamination procedures may be applied at this stage.
3. Dismantling and demolition of each plant and support buildings.

A preliminary assessment of the potential volumes of wastes produced by the D&D of O-1 and O-2 has been made (Larsson *et al.*, 2013; Bergh *et al.*, 2014). If the wastes from O-1 and O-2 are combined, it was estimated that a total of 1,900 tonnes of radioactive metal yielding greater than 10^6 Bq/kg would require shielding. Similarly, of total of 700 tonnes of radioactive concrete would require shielding. The potential efficacy of chemical and physical decontamination to reduce these amounts has yet to be considered.

The disposal options for the D&D wastes are also under study. The Final Storage For Reactor Waste (SFR) facility is a geological repository for short-lived, low- and intermediate-level radioactive waste (see Chapter 6). There are plans to expand the underground capacity of the SFR facility to accommodate D&D wastes (Rannemalm *et al.*, 2016). Interim storage of D&D wastes in temporary near-surface facilities may be required before the SFR facility is ready. Additional D&D information about the entire Oskarshamn Power Plant can be found in Larsson *et al.* (2013).

7.6 Review Questions

1. Give three reasons why nuclear facilities are removed from active service.
2. Define Safe Enclosure or DECON. Explain its advantage over Immediate Dismantling.
3. Define loose and fixed surface contamination.
4. Give three potential benefits of decontaminating surfaces.
5. How do strong acids decontaminate metal surfaces?
6. How to chelation agents help remove contaminants from metal surfaces?
7. Explain how the CORD procedure decontaminates surfaces.
8. How do strippable coatings remove contaminants from surfaces?
9. Explain how grinders and scabblers decontaminate concrete.
10. What type of radionuclides can be expected in a concrete bioshield?
11. What is the purpose of an Independent Spent Fuel Storage Installation?
12. Calculate Preliminary Remediation Goals (PRG) using a web-based, US EPA model for radionuclides in the Outdoor Worker scenario. Evaluate PRGs for soil. The Outdoor Worker is exposed for 225 days for 8 hours per day for 25 years. It was envisioned that the employee was a heavy machine operator who worked at West Valley (US) and was exposed to radiation from contaminated soil. The exposure pathways are external exposure to radiation, inhalation, and ingestion of blowing soil particles.

Go to
http://epa-prgs.ornl.gov/cgi-bin/radionuclides/rprg_search
Select Scenario: "Outdoor Worker"
Select Media: "Soil"
Select PRG type: "Defaults"
Select Units: "pCi"
Select three of any of the hundreds of radionuclides shown in the window.
Hit retrieve
Below Default, hit "Output to PDF"

Scroll down to the second line of the second table.

Answer the following questions:

a. For incidental ingestion of soil particles, which PRG for your three radionuclides is the most restrictive (the smallest PRG)?
b. For external exposure, which PRG for your three radionuclides is the most restrictive (the smallest PRG)?

Bibliography

Bergh, N., Hedin, G., and de la Gardie, F. (2014). Detailed Decommissioning Plans For The Oskarshamn And Forsmark Nuclear Power Plants in Sweden. *Waste Management Conference*, Phoenix, Arizona, March 2–6, 2014.

Bradley, D. J. (1997). *Behind the Nuclear Curtain: Radioactive Waste Management in the Former Soviet Union.* Battelle Press, Columbus Ohio.

Bushart, S. and Wood, C. (2003). The EPRI DFDX chemical decontamination process. *Waste Management Conference, Tucson, Arizona, Feb. 23–27.*

Chen, L., Chamberlain, D. B., Conner, C., and Vandegrift, G. F. (1997). *A Survey of Decontamination Processes Applicable to DOE Nuclear Facilities.* Argonne National Laboratory (Report number ANL-97/19).

Draine, A. (2009). *Decontamination of Medical Radioisotopes from Hard Surfaces Using Peelable Polymer-Based Decontamination Agents.* Master of Science Thesis, Colorado State University.

EPRI. (2004). *Pilot Demonstration of the EPRI DFDX Chemical Decontamination Technology.* Electric Power Research Institute, Palo Alto, California (Report number 1009572).

EPRI. (2006). *Connecticut Yankee Decommissioning Experience Report.* Electric Power Research Institute, Palo Alto, California (Report number 1013511).

GAO. (2006). *Nuclear Cleanup of Rocky Flats. DOE Can Use Lessons Learned to Improve Oversight of Other Sites' Cleanup Activities.* United States Government Accountability Office (Report number GAO-06-352).

Gurau, D. and Deju, R. (2015). The Use of Chemical Gel For Decontamination During Decommissioning Of Nuclear Facilities. *Radiation Physics and Chemistry*, 106, pp. 371–375.

IAEA. (1999). *State of the Art Technology for Decontamination and Dismantling of Nuclear Facilities.* International Atomic Energy Agency, Vienna (Technical Reports Series number 395).

ITRC. (2008). *Decontamination and Decommissioning of Radiologically Contaminated Facilities.* The Interstate Technology and Regulatory Council. Radionuclides Team. Washington, D.C.

Kinnunen, P. (2008). *ANTIOXI — Decontamination techniques For Activity Removal in Nuclear Environments.* VTT Technical Research Centre of Finland (Report number R-00299-08).

Larsson, H., Anunti, Å., and Edelborg, M. (2013). *Decommissioning Study of Oskarshamn NPP.* Swedish Nuclear Fuel and Waste Management Company (Report number SKB R-13-04).

NEA. (2011). *Decontamination and Dismantling of Radioactive Concrete Structures.* Nuclear Energy Agency (France), Organisation for Economic Co-Operation and Development (Report number NEA/RWM/R(2011)1).

Noynaert, L. (2012). Decontamination Processes And Technologies In Nuclear Decommissioning Projects. in Laria, M. (ed.), *Nuclear Decommissioning Planning, Execution, and International Experience.* Woodhead Publishing, Series in Energy, Sawston, UK, pp. 319–345.

Ponnet, M., Klein, M., and Rahier, A. (2000). Chemical Decontamination MEDOC Using Cerium IV and Ozone. *Waste Management Conference*, Tucson, Arizona, February 27–March 2, 2000.

Ponnet, M., Klein, M., Massaut, V., Davain, H., and Aleton, G. (2003). Thorough Chemical Decontamination With The MEDOC Process: Batch Treatment of Dismantled Pieces or Loop Treatment of Large Components Such As BR3 Steam Generator And Pressurizer. In *Waste Management Conference*, Tucson, Arizona, February 23–27, 2003.

Rahman, A. (2008). *Decommissioning and Radioactive Waste Management.* Whittles Publishing, Dunbeath, Scotland, UK.

Rannemalm, T., Eriksson, J., and Bergh, N. (2016). Decommissioning Planning for Nuclear Units at the Oskarshamn Site. In *PREDEC 2016: International Symposium on Preparation for Decommissioning. Organization for Economic Co-Operation and Development*, Lyon, France, February 16–18, 2016.

Roy, W. R. (2005). Iron. In Lehr, J. H. and Keeley, J. (eds.), *Water Encyclopedia: Oceanography; Meteorology; Physics and Chemistry; Water Law; and Water History, Art, and Culture.* John Wiley & Sons, Ostrander, Ohio, pp. 496–499.

Roy, W. R. and Holm, R. L. (2013). Decommissioning the University of Illinois Nuclear Research Laboratory. *Journal of Nuclear Energy Science and Power Generation Technology* (http://dx.doi.org/10.4172/2325-9809.S1-004)

Saling, J. H. and Fentiman, A. W. (2001). *Radioactive Waste Management*, 2nd Ed., Taylor and Francis, New York.

SKB. (2005). *Decommissioning of Nuclear Power Plants.* Swedish Nuclear Fuel and Waste Management Company (Report number Art308).

Sutton, M., Fischer, R. P., Thoet, M. M., O'Neill, M., and Edgington, G. (2008). *Plutonium Decontamination Using CBI DeconGel 1101 in Highly Contaminated Areas at LLNL.* Lawrence Livermore National Laboratory (Report number LLNL-TR-404723).

Taboas, A. L., Moghissi, A. A., and LaGuardia, T. S. (eds). (2004). *The Decommissioning Handbook.* American Society of Mechanical Engineers. Three Park Avenue, New York.

US DOE. (1998). *Concrete Shaver.* Innovative Technology, U.S. Department of Energy (Report number 1950).

US DOE. (2000). *ALARA*^{TM} 1146. *Strippable Coating.* Innovative Technology, U.S. Department of Energy (Report number DOE/EM-0533).

US DOE. (2017). *Hallam, Nebraska, Decommissioned Reactor Site.* U.S. Department of Energy. Legacy Management. Fact Sheet.

US DOE. (2018). *Site Management Guide.* U.S. Department of Energy. Legacy Management (LM Control Number: Guide-3-20.0-1.0-20.1).

US EPA. (2006). *Technology Reference Guide for Radiologically Contaminated Surfaces.* US Environmental Protection Agency (Report number EPA-402-R-06-003).

US EPA. (2011). *CBI Polymers DeconGel® 1101 and 1108 for Radiological Decontamination. Technology Evaluation Report.* U.S. Environmental Protection Agency (Report number EPA 600/R-11/084).

Varga, K., Baradlai, P., Hirschenberg, G., Nemeth, Z., Oravetz, D., Schunk, J., and Tilky, P. (2001). Corrosion Behavior Of Stainless-Steel Surfaces Formed Upon Chemical Decontamination. *Electrochimica Acta*, 46, pp. 3783–3790.

WNR. (2018). Decommissioning Nuclear Facilities. World Nuclear Association. Available at: http://www.world-nuclear.org/ [Accessed 10 February 2020].

Chapter 8

Transportation of Radioactive Materials

"What? Give up just as we are on the verge of success? Never!"

— Prof. Hardwigg in *Journey to the Center of the Earth* by Jules Verne

8.1 Introduction

The transportation of radioactive materials and wastes is an integral part of waste management (Salin and Fentiman, 2002). Radioactive wastes must almost always be moved from the location where they are generated to another location for interim storage or disposal. In the US, about 3 million shipments of radioactive materials are made each year (ANS, 2017). About 12,000 shipments of transuranic (TRU) wastes have been transported to the Waste Isolation Pilot Plant (WIPP) in New Mexico (see Chapter 6), and about 4,000 shipments of spent nuclear fuel have been conducted by road and rail since 1964 in the US.

Transportation must be accomplished as safely as possible to minimize human exposure to radiation, and to reduce potential barriers to the public's acceptance of nuclear energy. Safe transportation of radioactive materials must follow well-conceived procedures and planning coupled with proven engineered practices and experiences. Depending on the type of material, quantity, distance, and resources available, radioactive materials may be transported by road, rail, river, sea, and by air.

The transportation of radioactive materials is thoroughly regulated both in the US and internationally. The most comprehensive publication available to the reader on transporting radioactive materials are the guidelines provided by the International Atomic Energy Agency (IAEA) (IAEA, 2018). Shipments in the US are regulated by the US Department of Transportation and the US Nuclear Regulatory Commission using protocols that are consistent with the IAEA standards (ANS, 2017).

The types of radioactive materials and wastes that are routinely transported range from wastes generated by the uranium fuel cycle, decommissioning wastes, and medical, industrial, and research sources of radiation. The scope of this chapter is to introduce the subject of transporting radioactive materials with an emphasis on the packages used in shipping. It is beyond the scope of the chapter to paraphrase material that is readily available to the reader or to condense the voluminous requirements and regulations that are in place for each country that currently transports radioactive materials.

8.2 Radiological Characterization of Waste Packages

Both the selection of the type of waste package and the protocols needed to transport the material depend on the radiological characteristics of the contents. The greater the potential radiological hazards, the more robust and stringent are the package safety requirements (ANS, 2017). The requirements cover the type of package, labeling protocols, radiation shielding, loading and unloading during transport, storage, transportation routes, and the modes of transport.

8.2.1 *Low-Specific Activity (LSA)*

LSA can be defined as unshielded material that emits a relatively limited specific activity of radiation from sources that are essential uniformly distributed within the material (IAEA, 2018). In general, LSA materials pose a relatively small hazard because of their radioactivity. The LSA classification is subdivided into three groups that are based on the type of material, the physical form of the material, and the radionuclide-specific activity value: A_1 (special form) and A_2 (regular form). Special form is a solid with limited dispersibility or material inside a sealed capsule that is not in a powdered form. The A_1 term is the maximum activity of special form radioactive material permitted in a Type A package (discussed in the next section). The A_2 term is the maximum activity of radioactive material permitted in a Type A package that is not in a special form and is more restrictive than A_1. Examples of the activity values for selected radionuclides are given in Table 8.1. The three LSA groups are discussed in detail in IAEA (2018), but Table 8.2 provides examples of the progression from LSA I to III.

8.2.2 *Surface Contaminated Object (SCO)*

An SCO is a solid material, but as the name implies radionuclides are present on the surfaces but are not present at depth within the

Table 8.1. Examples of activity limits for selected radionuclides for Type A packages (from IAEA, 2018).

Radionuclide	A_1 (TBq) [Ci] for special form	A_2 (TBq) [Ci] for normal form
Americium-241	10 [270]	0.001 [0.03]
Nickel-63	40 [1,081]	30 [811]
Neptunium-237	20 [541]	0.002 [0.05]
Polonium-210	40 [1,081]	0.02 [0.54]
Uranium-235	Unlimited	Unlimited
Unknown β or γ sources	0.1 [2.70]	0.02 [0.54]

Table 8.2. Examples of the three LSA groups (from IAEA, 2018).

Low-specific activity group	Examples and partial characteristics
LSA-I	Uranium and thorium ores, natural and depleted uranium. Radioactive materials with an unlimited A_2.
LSA-II	Water with tritium ≤ 0.8 TBq/L (21.6 Ci/L). Materials with an estimated specific activity that does not exceed $10^{-4} A_2$/g of solids, and $10^{-5} A_2$/g for liquids.
LSA-III	Solid materials with an estimated specific activity that does not exceed $2 \times 10^{-3} A_2$/g of material.

Table 8.3. Examples of the three SCO groups (from IAEA, 2018).

Surface-contaminated object group	Partial characteristics
SCO-I	Non-fixed contamination on accessible surfaces that does not exceed 0.4 Bq/cm^2 (1.08×10^{-5} μCi/cm^2) for beta and gamma sources.
SCO-II	Fixed contamination on accessible surfaces that does not exceed 8×10^5 Bq/cm^2 (21.6 μCi/cm^2) for beta and gamma sources.
SCO-III	Non-fixed plus fixed contaminants on a large object that do not exceed 8×10^5 Bq/cm^2 (21.6 μCi/cm^2) for beta and gamma sources.

material. There are three categories of SCO (Table 8.3). The criteria for the three categories depend on whether the contamination is fixed or non-fixed (see Chapter 7), the type of surfaces (defined as accessible and inaccessible), and the activity and source of the radiation. The three SCO groups are discussed in detail in IAEA (2018), but Table 8.3 provides examples of the progression from SCO-I to III.

8.2.3 *Transport Index (TI)*

The Transport Index (TI) is a dimensionless shipping number for labelling packages, overpack, and freight containers. It is intended to communicate to transport crews and first responders the degree of potential radiological hazard in the event of an incident. The maximum dose rate at a distance of 1 m from the external surfaces of the package is determined by direct measurement. The measured dose as mSv/h is multiplied by 100.[1] If the TI is greater than 10 (Table 8.4), then the package or overpack must be transported by exclusive use meaning that it must be shipped alone without any other radiation sources.

[1]The origin of the 100 term is that 1 Sv = 100 rem. If TI = 0.8, for example, the dose rate should not be more than 0.8 mrem/h or 0.008 mSv/h × 100 = 0.8.

Table 8.4. Relationship between the maximum dose rate at 1 m from external surfaces of a package, the Transportation Index (TI), and the transportation category (from IAEA, 2018).

Maximum dose rate (mSv/h)	TI (unitless)	Transportation category
≤0.005	0	I — White
>0.005 but ≤0.5	>0 but ≤1	II — Yellow
>0.5 but ≤2.0	>1 but ≤10	III — Yellow
>2.0 but ≤10	>10	III — Yellow[2]

8.2.4 *Critically Safety Index (CSI)*

The Critically Safety Index (CSI) is a shipping package characterization that is applied to the shipping of fissile materials. The purpose of the CSI is to reduce the probability of a self-sustained nuclear reaction. It is beyond the scope of this chapter to describe how a CSI is determined. A CSI depends on the type of fissile material, mass, shape, volume, density, and the presence of neutron absorbers, moderators, and reflectors. The determination is made by the application of computer models applied to the specific material and the shipping requirements. IAEA (2018) provided guidelines and recommendations for the determination of a CSI.

8.3 Shipping Packages

The purpose of all shipping packages for radioactive materials is to provide adequate shielding from radioactive materials during routine handling and during accidents when transported. There are five different categories of shipping packages:

8.3.1 *Excepted Package*

This type of container is used for radioactive material exhibiting radiation levels that are regarded as insignificant. Excepted packages are

[2]This category is a candidate for exclusive use shipping in which the package is transported alone.

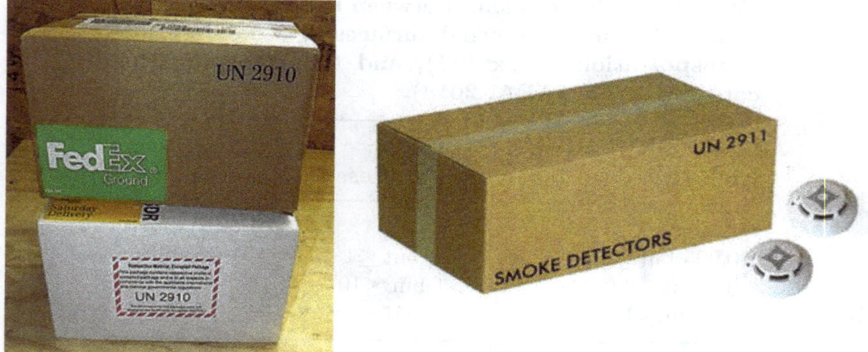

Figure 8.1. Examples of excepted packages (US DOT, 2008).

not required to be tested or designed to survive any transportation accidents. It assumed that all of the contents could be released during an accident (US DOT, 2008). An example of an excepted package could be a cardboard box (Fig. 8.1) used to ship an ionizing smoke detector that typically contains a small amount of americium-241.

8.3.2 *Industrial Package*

This type of container is used to transport relatively low-level radioactive material such as LSA materials. There are three categories of Industrial Packages (IP): IP-1, IP-2, and IP-3. An IP-1 package has no testing requirements for normal transport conditions. An IP-2 package must survive a free drop and stacking requirement. The IP-3 package has an additional penetration or compression test requirement. The survival requirements of an IP-3 package are identical to those for a Type A package (US DOT, 2008) which is discussed in the next section.

Commonly used Industrial Packages are steel drums and boxes. The relatively familiar 200-L drum (or the 55-gallon drum in the US, and the 44-gallon drum in the UK) are often used to ship low- and intermediate-level radioactive wastes (see for example Droste, 2015). IAEA (1993) provided an interesting history about the designs and uses of various materials (steel, cast iron, plastics, and concrete) to

Figure 8.2. Industrial Packages (IP-2) for radioactive wastes in France (Socomelu, 2020). Available at: https://www.socomelu.com [Accessed 10 February 2020]. Used with permission by SOCOMELU.

produce waste containers for low- and intermediate-level radioactive wastes.

8.3.3 *The Type A Package*

This package is used to transport a relatively small but significant source of radiation. It is assumed that a Type A package will maintain its physical integrity under conditions of normal transport and during rough handling or minor accidents. A Type A package may be damaged in a *severe* accident resulting in the release of some of its contents (US DOT, 2008). Solid materials in Type A packages are required to survive four tests that are intended to simulate "rough handling" during normal transportation (Table 8.5). Depending on the performance standards required, Type A packages are diverse. They can be made of steel, wood, or fiberboard (Fig. 8.3). Type A containers occur as instrument carrying cases, small steel cans for isotopes, steel drums, and relatively large steel shipping casks.

Table 8.5. Endurance tests required of Type A packages (selected from IAEA, 2018).

Test	Intended scenario
Water Spray 5 cm/h for at least 1 h	Sitting uncovered on a loading dock or road exposed to heavy rain.
Free Drop Package is dropped 0.3–1.2 m (depending on its weight) onto a hard surface.	Falling off a vehicle or loading dock during loading/unloading or an accident.
Stacking A weight 5 times heavier than the package is placed on the package for 24 h.	Being crushed at the bottom of a stack of heavier packages.
Penetration A 6 kg bar is dropped 1 m onto the upper surface of the package.	Being struck by a loose object or by heavy freight with a sharp corner during loading/unloading or an accident.

8.3.4 *The Type B Package*

This package is designed to transport material with the greatest radioactivity. These packages are designed to survive *severe* accidents. Type B packages must pass all the Type A tests (Table 8.5) but are also required to endure additional tests that are intended to simulate worst-case scenarios such as a train derailment with a subsequent fire (Table 8.6).

Type B packages are diverse and can vary from hand-held instruments to heavily shielded steel casks (Fig. 8.4). Type B shipping casks are used to transport transuranic (TRU) wastes (see Chapter 5) from US DOE sources to the Waste Isolation Pilot Plant in New Mexico (see Chapter 6). The Transuranic Package-Transporter Model 2 (TRUPACT-II) and the HalfPACT casks are used to transport Contact-Handled TRU wastes (Fig. 8.5). Both casks consist of an inner and outer stainless-steel cylinder that are separated by polyurethane foam that serves as an impact limiter. The payload volume is also protected by an upper and lower layer of a ceramic

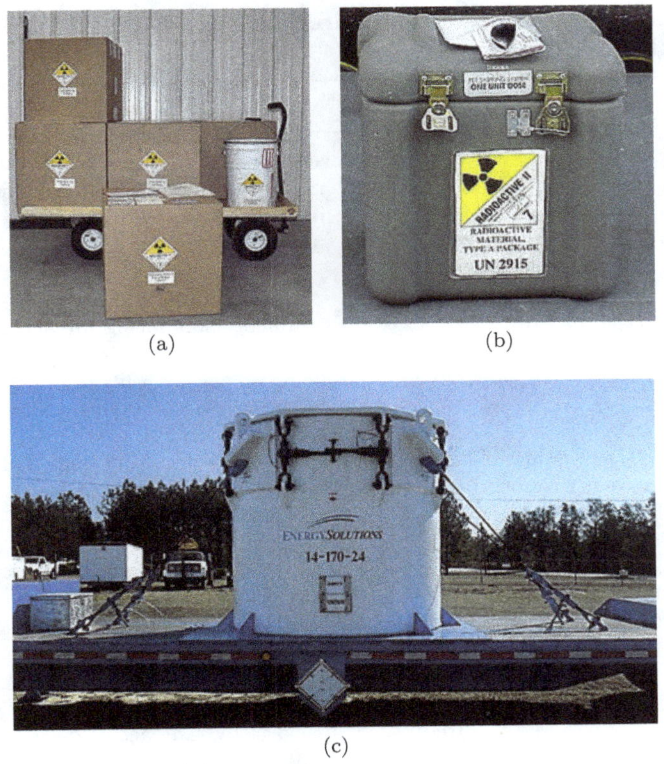

(a) (b)

(c)

Figure 8.3. Type A Packages (a and b) US DOT, (2008), and (c) Energy*Solutions* 14-170 Cask (Energy*Solutions*, 2020). Available at: https://www.energysolutions.com [Accessed 12 February 2020]. Used with permission by Energy*Solutions*.

fiber donnage. The TRUPACT can hold fourteen 55-gallon waste drums. The HalfPACT is used to transport seven waste drums and has the advantage of being lighter for road transportation.

The RH-72B shipping cask is used to transport remote-handled (RH) TRU wastes (Fig. 8.6). It consists of a relatively large steel cylinder that is capped on both ends by circular impact limiters that contain polyurethane foam. Sandwiched in between layers of stainless steel, the cylinder has 1.875-inch (48 mm) liner made of lead to shield workers from the gamma radiation inherent to Contact-Handled TRU wastes.

Table 8.6. Endurance tests required of Type B packages[3] (selected from IAEA, 2018).

Test	Intended scenario
Free Drop I The package is dropped 9 m onto an unyielding surface such that the weakest surface (such as a corner) impacts first.	Presumedly simulating impact during a major accident such as a traffic or rail incident.
Free Drop II (Puncture) The package is dropped 1 m onto a mounted 15-cm diameter, 20-cm long steel rod.	Presumedly simulating impact during an accident against a sharp corner or being struck by a high-velocity loose object during a major accident.
Free Drop III (Dynamic Crush) A 500-kg mass is dropped 9 m onto the package.	Presumedly simulating impact during an accident possibly with other freight packages present.
Thermal (engulfing flames) The package is subjected to at least 800°C for at least 30 minutes.	An accident results in the ignition of flammable liquids such as diesel fuel or flammable chemicals in railroad tank cars.
Thermal (solar insolation) The package is exposed to direct sunlight at 38°C.	The generation of heat from within the package by radioactive decay contributes excessively to the overall temperature of the package.
Water Immersion The package is submerged in water that is at least 15 m deep for at least 8 h.	The package is separated from the mode of transportation and falls into a river, lake, or the sea.

Type B shipping casks for spent nuclear fuel are the transport and dual-use casks discussed in detail in Chapter 6. For example, Holtec's HI-STAR (Storage, Transport, And Repository) 80 cask is a dual-purpose cask in the sense that it can be used for interim

[3]The testing of packages may be conducted using full-scale prototypes, and by using scale models. One-quarter scale models are often used in engineering tests (National Research Council, 2006). In addition to experimental data, numerical simulations are used during development and testing (Droste, 2015).

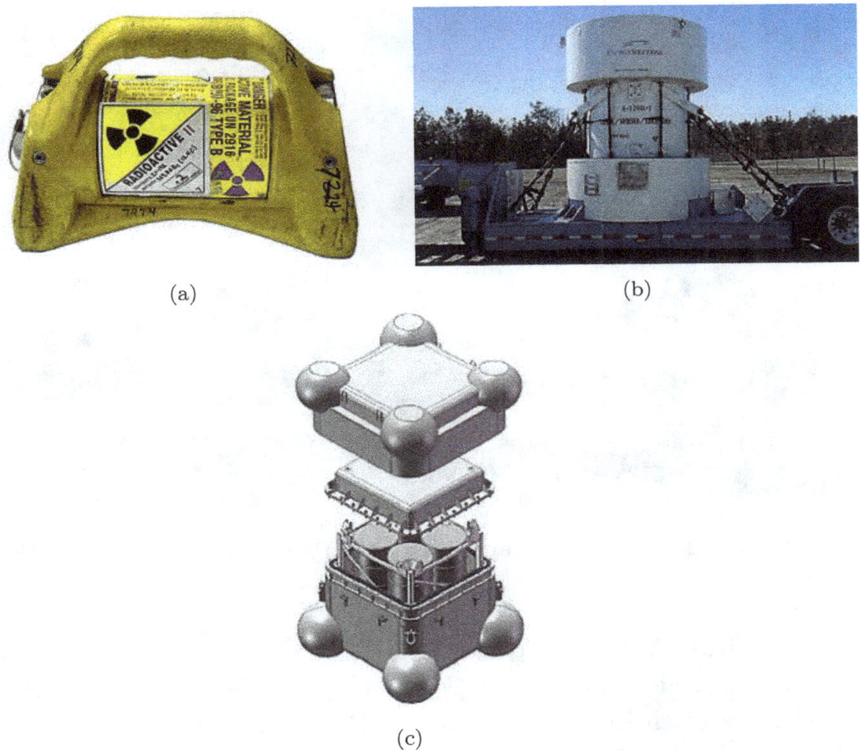

Figure 8.4. Type B casks: (a) Radiography camera (REMM, 2020). Available at: https://www.remm.nlm.gov [Accessed 12 February 2020]. (b) Energy*Solutions* 8-120B-1 cask (Energy*Solutions*, 2020) Available at: https://www.energysolutions.com [Accessed 12 February 2020] (c) Robatel R77s cask (Robatel Technologies, 2015). Available at: http://www.robateltech.com [Accessed 12 February 2020]. Used with permission by Robatel Technologies and Energy*Solutions*.

storage at an Independent Facility for Spent Nuclear Fuel, then later for transport to a geological repository.

8.3.5 *The Type C Package*

It is envisioned that the Type C package will be used for air transport of highly radioactive material. Type C packages would need to meet the durability tests for Type A and B packages, and some additional tests. This type of package has not yet been fully developed

Figure 8.5. The HalfPack and two TRUPACT-II casks (US DOE, 2016).

(IAEA, 2018). A Type C package, however, has been developed in the Russian Federation and used to transport spent fuel during the Russian Research Reactor Fuel Return Program (Budu *et al.*, 2014). The TUK-145/C design consists of a Czech SKODA VPVR/M cask placed into an Energy Absorption Container (EAC). The EAC is a titanium cylinder in which about 2,000 titanium alloy spheres are packed around the inner cask (Fig. 8.7). The TUK-145/C was approved by Russia, and was used to transport spent fuel from Romania, Libya, and Uzbekistan.

The PMATP, a Type C Prototype Package

Sandia National Laboratories designed and tested a prototype package for air transport of plutonium (Pierce *et al.*, 2003). It was designed to transport 7.6 kg of plutonium and to survive a "worst case" airplane crash. The package is called the Perforated Metal Air Transportable Package (PMATP). The design consists of the plutonium being placed in a stainless steel can with is then packed into an overpack composed of layers of aluminum and heat-resistant cloth layers. A stainless-steel shell encases the overpack.

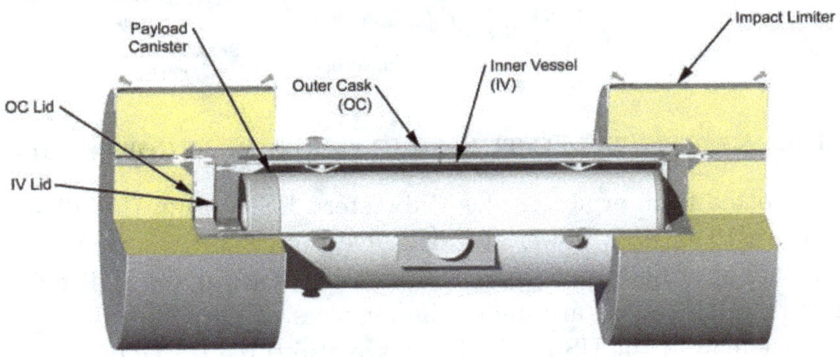

Figure 8.6. The RH TRU 72B Type B transport cask (US NRC, 2009). Available at: https://www.nrc.gov/docs/ML0916/ML091630341.pdf [Accessed 12 February 2020].

8.4 Transportation Planning in the Midwestern US

The Midwestern US[4] was chosen as a source of material for this section because of the number of radioactive wastes that have been

[4]The Midwestern US is defined as the large area containing Illinois, Indiana, Iowa, Kansas, Michigan, Minnesota, Missouri, Nebraska, North Dakota, Ohio, South Dakota, and Wisconsin.

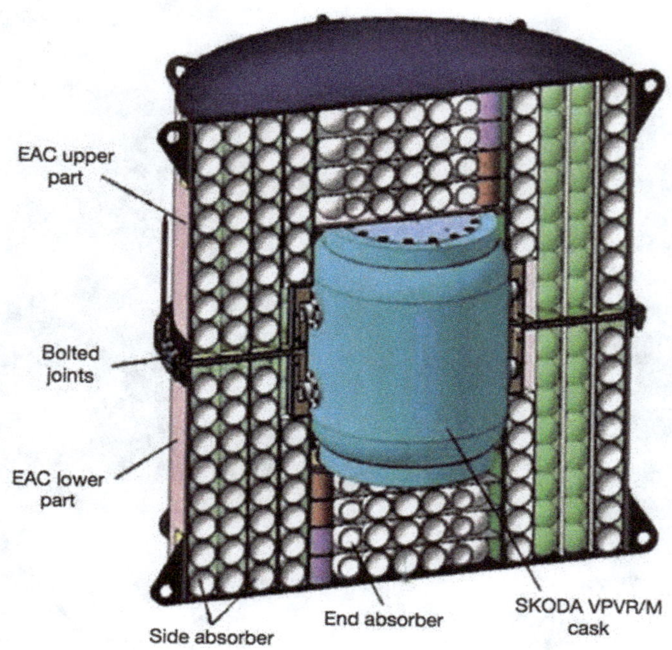

Figure 8.7. The Type C TUK-145/C package (from Cribier, 2014).

transported from or across the Midwestern US. Within this 12-state area are 19 nuclear power stations, five power stations that are in some stage of decommissioning, 6 research reactors, and 20 former or current sites that are under the auspices of the Office of Legacy Management of the US DOE. There are also three US DOE national research laboratories in the Midwest. Therefore, there are several potential sources of low-level and TRU waste, decommissioning wastes, and spent nuclear fuel. There is also the necessity to transport fresh nuclear fuel to the power and research reactors. There are no commercial or US DOE low-level radioactive waste disposal sites in the Midwest.

Consequently, the wastes must be transported from the Midwest to disposal sites in other parts of the country (see Chapter 4). Moreover, because the Midwest is located between the eastern and western parts of the US, radioactive waste shipments have often crossed the Midwest as they are transported to their destination.

The Midwest Radioactive Materials Transportation Committee (CSG, 2005, 2017) created a guidance document based on the experiences gained from numerous shipping campaigns. The scope of this guidance document is the transportation of spent nuclear fuel, TRU wastes, high-level radioactive wastes, and Highway Route Controlled Quantity (HRCQ) material.[5] Central to transportation planning is an acceptable Transportation Plan. According to CSG (2017), the plan should include:

1. The purpose of the shipment.
2. An emergency management section.
3. A communication section.
4. Details about the mode of shipping and the carriers to be used.
5. The preferred route of transport and alternative routes.
6. An incident/accident recovery section.
7. Points of contact during shipping.

The specific shipper should provide a Carrier Management Plan that contains:

1. An emergency response section.
2. Information about driver or rail crew training for radioactive substances.
3. Accident recovery protocols.
4. Security during transport.
5. Available equipment to conduct the shipping campaign.
6. A communication section.

CSG (2017) recommended that shippers should begin the transportation planning process 2 years before the anticipated commencement of the shipments to allow the Committee to review and provide input. All shippers in the Midwestern states must use US NRC-approved Type B packages for materials that are not

[5]The US DOT defined the HRCQ quantity as that in a single Type B package of 3,000 times A_1 of the radionuclide in special form, 3,000 times A_2 of a radionuclide in normal form, or 1,000 TBq (27,000 Ci) — whichever activity is the least.

low-level radiation materials. The State of Illinois requires a state escort for security for all shipments. Other requirements may include in-route inspections, and the designation of safe parking areas away from highly populated areas, hospitals, schools, residential areas, Interstate rest areas, and highway shoulders. Shipping materials during December and January are generally prohibited because of the likelihood of severe winter weather in the Midwest.

US DOE (2016) provided a detailed generic transportation plan for shipping TRU wastes to WIPP. Some of the major parts in the guidance document are as follows:

1. Roles and responsibilities.
2. Transportation system (route selection, shipping packages and papers, placards, the carrier contractor).
3. Operations (routing, notifications, tracking using TRANSCOM, road construction, weather delays, safe parking).
4. Security.
5. Communications.

This guide has the potential to serve as a template for other transportation plans. The reader should study this guide for content and ideas for future use.

8.5 Transportation Labeling

IAEA (2018) provided detailed requirements for labeling each shipping package, overpack, and freight (typically reusable) container. The labels and placards that are used are based on the transport categories given in Table 8.4 which are based on TI and surface dose rate (Fig. 8.8). Common to each placard is the number 7 which is the United Nations (UN) code for radioactive materials. Each placard is diamond shaped and is intended to be readily visible and durable. It is intended, for example, to communicate to first responders at an accident scene that radioactive materials are present. The placards are also to identify the specific radionuclide and activity within the package. A placard for fissile material also contains the criticality safety index. Lastly, there is an additional orange placard

Figure 8.8. Transportation labels for (left to right) Category I, II, and III packages (US DOT, 2008).

that contains UN numbers for each combination of package and radioactive material. For example, 3324 means Radioactive Material, Low specific activity (LSA-II). Fissile.

8.6 Risk Assessments in Transporting Radioactive Materials

All activities required to transport radioactive materials have some associated risks. Risks are the unwanted consequences of an event. Risks during transportation can be evaluated under two conditions: normal operations and accident conditions (Salin and Fentiman, 2002). The two populations that could be potentially exposed to radiation are the transport personnel and the general public along the route used for transport. Because of the rigor of testing coupled with required shielding endemic to transport packages, the potential for radiation exposure in normal and accident scenarios is relatively insignificant. Salin and Fentiman (2002) provided an excellent summary of risk assessments that were conducted in the 1980s. The National Research Council (2006) also conducted a detailed risk assessment of transporting radioactive materials under normal conditions and accidents including both road and rail modes of transport. The outcome of this analysis is the now classic book "Going the Distance?" Among the principal conclusions reached by the Council were:

1. There are no known technical barriers to the safe transport of spent nuclear fuel and high-level waste in the US.
2. Transportation packages provide a robust barrier to the release of spent fuel and high-level wastes shipments under normal and accident conditions.
3. Societal risks may include negative economic impacts and increased anxieties along the transportation route.

The Council made recommendations for the third conclusion about greater involvement of the public in the planning process. The Council also cautioned that terrorist attacks are another concern for transportation security. Note that since this observation was made, armed escorts during shipping campaigns are mandatory.

8.7 Transportation of Radioactive Materials in Europe

The French Republic is a world leader in the development of nuclear energy and waste management (see Chapter 10). It is also an excellent example of transportation in Europe. The Nuclear Safety Authority (of France) summarized that about 900,000 packages containing radioactive materials are transported each year in France (ASN, 2012). Of that amount, 11,000 shipments are related to the nuclear fuel cycle: fresh reactor fuel, MOX fuel, spent reactor fuel, and various shipments from other countries. About 60,000 Type B packages are transported each year in France.

Connolly and Pope (2016) provided an historical review of transporting spent nuclear fuel worldwide. From 1962 to 2016, at least 25,400 shipments were made by rail, road, river, and sea. These shipments amounted to about 87,000 tonnes of heavy metal. At least 130 Type B casks of vitrified waste were transported from France to the country where the used nuclear fuel originated.

From 2007 to 2013, about 50 shipping campaigns were completed under the Russian Reactor Fuel Return Program. About 25 countries participated. One hundred Type B shipping casks were used in the program These casks were the Russian-made TUK 19 cask

Figure 8.9. The SKODA VPVR/M Type B cask [Available at https://www.iaea.org/newscenter/multimedia/. [Accessed 11 October, 2020]. Used with permission of the IAEA.

and the Czech Skoda VPVR/M cask (Fig. 8.9). Connolly and Pope (2016) concluded that the transportation of spent nuclear fuel has been accomplished routinely and safely in many countries for decades.

Case Study: The transport of radioactive materials between France and Germany

Alter (in ASN, 2012) provided a synapsis of the movement of radioactive materials across the French–German border during the last 40 years. These materials have included spent nuclear fuel, vitrified high-level wastes, plutonium, MOX fuel elements, uranium hexafluoride, fresh nuclear fuel, and isotopes. The modes of transport were by road and rail. Only low- and intermediate-level wastes have not been shipped across the border. France reprocessed used nuclear fuel for Germany on the condition that the byproduct waste, once vitrified in France, is to be returned to Germany for disposal. Alter (ASN, 2012) reported that during a 15-year period, reprocessing 5,309 tonnes of used fuel generated 108 Type B transport casks that were used to ship to Germany. The transport casks were TS 28 V, CASTOR HAW 28/20 CG, and TN 85 (see Droste, 2015 for details about the CASTOR HAW and TN 85 casks).

8.8 Marine Transportation of Radioactive Materials

In order to meet national requirements for re-fueling, Japan had to ship used nuclear fuel to France and the UK for chemical reprocessing and had to accept the vitrified high-level reprocessing waste for disposal. To meet this requirement, Pacific Nuclear Transport Limited (PNTL) has been engaged to ship used nuclear fuel, MOX fuel, and the vitrified wastes since 1969. About 180 sea shipments have been made using purpose-built ships that travel non-stop between Europe and Japan. Type B casks have been used such as the TN 28 VT and the TN 12/2 (PNTL, 2016). About 7,137 tonnes of used fuel have been shipped to Europe for reprocessing prior to 2007 (WNA, 2017). About 635 tonnes of vitrified waste have been shipped to Japan from France. Sea shipping between France and Japan is on-going. To date, there have been no incidents or lost shipments while in transport.

Marine transportation is an integral part of radioactive waste management in the Kingdom of Sweden. Sweden's situation is somewhat unique in that its three power plants are located along the coast of the country. Spent fuel and intermediate-level wastes are routinely transported by sea for disposal on land. The spent fuel is transported to Oskarshamn to be stored at the Central Interim Storage Facility for Spent Nuclear Fuel (see Chapters 6 and 10). The intermediate-level waste is transported by road and sea to the Final Repository for Short-Lived Radioactive waste (see Chapter 10). The transport casks that have been used are the TN 17/2, and were shipped by purpose-built cargo vessels: the m/s Sigyn (from 1982 to

2013) and the m/s Sigrid from 2013 to the present. About 50 spent fuel casks have been shipped by sea each year. To date, there have been no incidents or lost shipments at sea.

8.9 Review Questions

1. Define a low-specific activity material.
2. What is a special form material?
3. If the Transportation Index is greater than 10, how will the package be transported?
4. Give an example of an Exempted Package.
5. An Industrial Package 3 (IP-3) must pass the same testing requirements as ______________.
6. What type of shipping conditions must a Type B package survive?
7. Describe a TRUPACT-II Type B package. What is it used for?
8. Describe the Free Drop test for a Type A package.
9. What is the intended scenario of the Water Immersion test for a Type B package?
10. Describe the Type C TUK-145/C cask.
11. (a) Based on the information given in Tables 8.5 and 8.6, create some testing procedures for a new Type C cask. (b). Design a cask to meet those tests.
12. Describe the history of shipping radioactive materials across the French-German border.

Bibliography

ANS. (2017). *The Safety of Transporting Radioactive Materials.* Position Statement 18. American Nuclear Society (Report ANS-18-2017).

ASN. (2012). *The Safety of Transport of Radioactive Materials.* The Nuclear Safety Authority (France) (Contrôle Review 193).

Connolly, K. J. and Pope, R. B. (2016). *A Historical Review of the Safe Transport of Spent Nuclear Fuel.* U.S. Department of Energy (Report FCRD-NFST-2016-000474).

Cribier, M. (2014). ^{144}Ce source for CeSoX. Transportation Studied in France. CeSoX Meeting. February 5–7, 2014 (Accessed on-line Oct. 5, 2020) [Available at slideserve.com].

CSG. (2005). *Handbook of Radioactive Waste Transportation.* Council of State Governments, Midwest Office, and the Midwestern Radioactive Materials Transport Committee.

CSG. (2017). *Planning Guide for the Shipments of Radioactive Materials through Midwestern States.* Council of State Governments, Midwest Office, and the Midwestern Radioactive Materials Transport Committee.

Droste, B. (2015). Packaging, Transport, and Storage of High-, Intermediate-, and Low-Level Radioactive Wastes. In Sorenson, K. B. (ed.), *Safe and Secure Transport of Radioactive Materials*, Chapter 15, Woodhead Publishing, Sawston, UK.

IAEA. (1993). *Containers for Packaging of Solid and Intermediate Level Radioactive Wastes.* International Atomic Energy Agency (Technical Reports Series number 355).

IAEA. (2018). *Regulations for the Safe Transport of Radioactive Material.* International Atomic Energy Agency (Specific Safety Requirements number SSR-6 (Rev. 1)).

National Research Council. (2006). *Going the Distance?* The National Academies Press, Washington, DC.

Pierce, J. D., Gronewald, P., Mould, J., and Oneto, R. (2003). *Radiant Heat Test of Perforated Metal Air Transportable Package (PMATP).* Sandia National Laboratories (Report number SAND2003-2750).

PNTL. (2016). *Sea Shipments of Radioactive Waste from Europe to Japan.* Pacific Nuclear Transport Limited (Fact Sheet).

Saling, J. H. and Fentiman, A. W. (2001). *Radioactive Waste Management,* 2nd Ed., Taylor and Francis, New York, NY.

US DOE. (2016). *TRU Waste Transportation Plan. Revision 4.* US Department of Energy (Document DOE/CBFO-98-3103).

US DOT. (2008). *Radioactive Material. Regulations Review.* U.S. Department of Transportation.

WNA. (2017). *Transport of Radioactive Materials.* World Nuclear Association. Available at: http://www.world-nuclear.org/ [Accessed 15 February 2020].

Chapter 9

Environmental Restoration in the United States

"I have an idea, my dear boy; it is none other than this simple fact; we shall not come out by the same opening as that by which we entered."

— Prof. Von Hardwigg in *Journey to the Center of the Earth* by Jules Verne

9.1 Introduction

In this chapter, the term "environmental restoration" refers to the characterization and remediation of industrial facilities that were part of the Nuclear Weapons Complex. During World War II and the Cold War, the US Federal government developed and operated a vast network of industrial facilities for research, production, and testing of nuclear weapons. This national effort required uranium mining, milling, fuel and target fabrication, plutonium production and separation, weapons design, testing, and finally weapon assembly (US DOE, 2019a).

The Complex produced more than 70,000 nuclear warheads. In 1989, the Complex consisted of about 107 sites in 35 states and covered an area of 8,094 km^2 (3,125 $miles^2$) (NGA, 2012) with the largest facilities in Idaho, Nevada, South Carolina, and Washington. With the end of the Cold War in about 1991, and the subsequent collapse of the Soviet Union, the need for the Complex was greatly reduced, and many facilities were shut down. While World War III did not occur, almost every site in the Nuclear Weapons Complex was left contaminated with radioactive or hazardous materials because waste management was not a major priority. This residual contamination was diverse in form, concentration, and location including large volumes of uranium mill tailings, abandoned buildings and laboratories, equipment and tools, below-ground waste tanks, waste disposal pits and landfills, and in some cases, multiple plumes of contaminated groundwater. Legacy wastes also included chemical and radioactive waste mixtures derived from the plutonium extraction process and spent nuclear fuel that was abandoned when the production of nuclear weapons ceased. US DOE (1997) estimated that the volume of radioactive wastes related to nuclear weapons at 49 sites was 24,000,000 m^3, and that the corresponding radioactivity was 33,000,000 TBq (900,000,000 Ci). The vast majority of the contamination was identified in soil and groundwater.

The US DOE is responsible for the legacy wastes created by the Nuclear Weapons Complex. In 1992, the Federal Facility Compliance Act (FFCA) made it possible for the impacted areas to work more closely with US DOE in implementing cleanup and waste disposal strategies. The Office of Legacy Management was established in 2003 and it is responsible for ensuring that US DOE's post-closure responsibilities are met. The on-going remediation effort has been called the largest environmental cleanup program in the world (NGA, 2012). By 2012, the size of the Complex had been reduced to 824 km^2 (318 $miles^2$) at 17 sites in 11 US States (NGA, 2012). Current estimates (NGA, 2012) indicate that about 50 more years are needed to complete site cleanup of the entire Complex.

It is, however, important to understand the scope of the cleanup effort. US DOE acknowledges that some legacy sites cannot be sufficiently cleaned up to be released to the public for unrestricted use because of technical and financial limitations (US DOE, 2019a). In situations where unrestricted land use is not possible, US DOE's programs provide long-term monitoring of the potential movement of residual radionuclides, and institutional controls that limit human exposure for decades, and at some sites, in perpetuity (NGA, 2012). The case studies presented in this chapter are intended to provide diverse examples of different types of sites that were part of the Nuclear Weapons Complex, and to illustrate the approaches used to reduce their threat to human health and the environment.

9.2 The Grand Junction (Colorado) Sites

There were two different uranium recovery facilities in the Grand Junction area in arid, western Colorado. One was located southwest of the city of Grand Junction, next to the Gunnison River. The second facility was south of Grand Junction and was located along the Colorado River. In 1943, the US War Department acquired the 22-ha (54-acre) Grand Junction site to support the Manhattan Engineer District. A uranium refinery was operated from 1943 to 1946. Beginning in 1953, the US Atomic Energy Commission operated two pilot uranium mills, and about 27,200 tonnes (30,000 tons) of ore were processed before milling operations ceased in 1958 (US DOE, 2019b).

In 1950, the Climax Uranium Company began operating a uranium and vanadium mill at the Colorado River site. The 46-ha (114-acre) mill operated until 1970 and produced 2.0 million tonnes (2.2 million tons) of mill tailings (US DOE, 2019c). From 1950 to 1966, the mill tailings were made available to the general public and to contractors as construction fill, and as a component in concrete. After closure, the Climax Uranium Company demolished most of the buildings at the facility.

The milling operations during the decades of uranium and vanadium production resulted in soil and groundwater contamination. The technology chosen for the disposition of the mill tailings and contaminated soil was placement in a nearby disposal cell. A relatively unique feature at the Grand Junction Site is the presence of a 91-m (300-foot) deep, cased borehole that was used to dispose of radium foil. The removal of the tailings and contaminated soil began in 1989, and the construction of the Grand Junction Disposal Cell began a year later. The area of the disposal cell is 38 ha (94 acres) and was excavated into about 12 m (40 feet) of alluvium, colluvium,

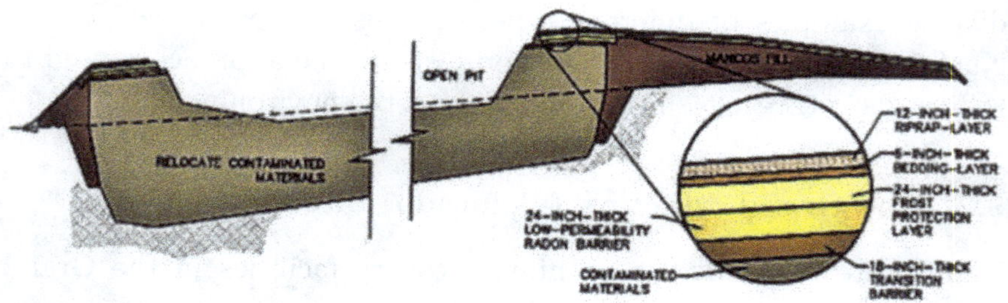

Figure 9.1. Cross-section of the Grand Junction Disposal Cell (US DOE, 2019c).

and terrace gravels, and the underlying Mancos Shale (Fig. 9.1). US DOE chose the site location based on:

Remoteness: Located about 20 km (18 miles) from the city of Grand Junction, the nearest resident is 3.2 km (2 miles) distant.

Geology: The shallow groundwater in the area is unacceptable for human consumption. Moreover, the occurrence of the thick, relatively water-impermeable shale below the disposal cell would help reduce the movement of leachate away from the disposal cell.

About 3,400,000 m^3 (4,400,000 yards^3) of tailings have been disposed (US DOE, 2019c). The disposal cell will remain open until it is filled or until the year 2023. Where the cell has been filled, the wastes are covered with multiple barriers (Fig. 9.1). The lowermost barrier is a layer of compacted clay-sized material that is intended to be a barrier to radon produced by radioactive decay, and to precipitation. Above that is a layer of coarse sand to facilitate drainage, then a fine-sand bedding layer to help secure an outer riprap layer that is used to protect the cell from erosion, and to discourage the establishment of volunteer plants.

The groundwater in the alluvial aquifer below both sites is contaminated. The contamination from the Colorado River site can be detected at 1,005 m (3,300 feet) downgradient from the site (US DOE, 2019c). Shallow groundwater samples collected in 1998

Table 9.1. Partial summary of 1998 groundwater data at the Grand Junction Site. The concentrations shown are the maximum values as mg/L (US DOE, 1999).

Constituent	Plume	Background
Ammonium (as NH_4)	233	0.32
Chloride	1,160	991
Molybdenum	0.30	0.12
Uranium	2.50	0.07
Vanadium	0.83	0.005

(US DOE, 1999) indicated that the solution concentrations of ammonium, chloride, molybdenum, uranium, and vanadium were greater than background levels (Table 9.1). However, US DOE chose a "no action alternative" for groundwater remediation. This chosen technology is now called *natural flushing*. In principle, this is a passive approach in which natural precipitation infiltrates the ground, then dilutes and displaces (flushes) contaminants from the unconsolidated materials and the shallow groundwater during relatively long periods of time — 50 to 80 years. Depending on the physicochemical form of the contaminant, the downward movement of shallow groundwater could also dissolve solid phases and desorb bound contaminants.

US DOE (1999) concluded that natural flushing would reduce site-related contaminants to near-background levels in less than 100 years. US DOE concluded that natural flushing was the appropriate approach because the off-site groundwater quality was poor and was not a current or potential source of drinking water. It also concluded that the no-action alternative would not result in any adverse health effects on the human population.

US DOE is committed to the long-term groundwater monitoring to document the progress of natural flushing. For example, seven monitoring wells were installed on-site and downgradient at the Grand Junction Site. The wells are being sampled annually, and the

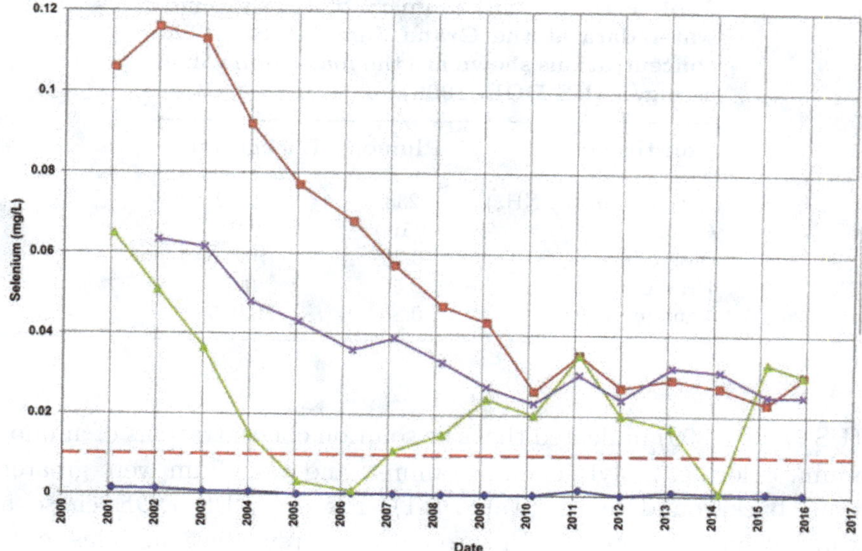

Figure 9.2. Selenium concentrations at four monitoring wells at the Grand Junction Site (US DOE, 2016). The dashed line is maximum contaminant level (0.01 mg/L).

water samples are chemically characterized. Results for a 15-year interval from 2001 to 2016 for the concentrations of manganese, molybdenum, selenium, sulfate, and uranium at the Grand Junction Site have been inconsistent. For example, the concentrations of dissolved selenium in three monitoring wells decreased during the 15-year interval (Fig. 9.2). In contrast, the solution concentrations of uranium in the same monitoring wells decreased in two monitoring weeks but were increasing after about 2014 in one well (Fig. 9.3). It seems likely that it is too soon to evaluate the long-term efficacy of natural flushing. Moreover, US DOE (2011a), cautioned that additional monitoring wells were needed at the Grand Junction Site, and that there is no established protocol for assessing when natural flushing is complete.

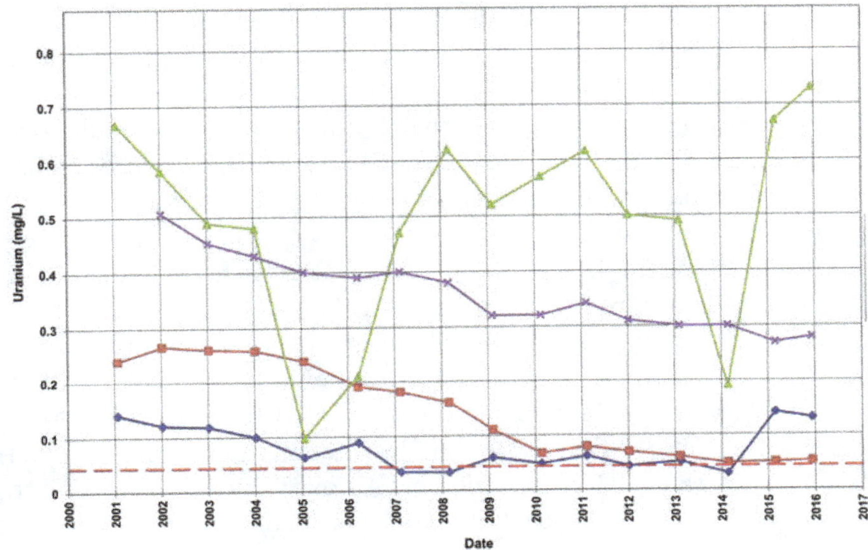

Figure 9.3. Uranium concentrations at the same four monitoring wells in Fig. 9.2 at the Grand Junction Site (US DOE, 2016). The dashed line is the maximum contaminant level (0.044 mg/L).

9.3 The Rifle (Colorado) Sites

There were two former uranium and vanadium recovery mills located near Rifle, in western Colorado. The Old Rifle mill operated from 1942 to 1958 by leaching uranium ore with sulfuric and hydrochloric acids. About 629,000 tonnes (693,495 tons) of ore were processed from 1947 to 1958 that created a 5.3-ha (13-acre) tailings pile (US DOE, 1998). The New Rifle mill was constructed in 1958 west of the city of Rifle. Both uranium and vanadium were produced from 1958 to 1972, generating about 1,727,000 m^3 (2,259,000 $yards^3$) of tailings. Surface remediation of the Rifle Sites began in 1992 and was completed in 1996.

Similar to the Grand Junction Site, the technology chosen for the final disposition of the mill tailings was a disposal cell that was constructed approximately 9.7 km (6 miles) north of the New Rifle Site. This disposal site was selected because it was located in a sparsely populated area, and because of the local geology. The site is underlain by a relatively water-impermeable sandstone-siltstone layer (US DOE 1997), and the groundwater is not a current or future source of drinking water because of poor water quality and low-well yield (US DOE, 2018a).

Tailings and tailings-contaminated materials from both former processing sites were relocated to the Rifle disposal site. The Rifle Disposal Cell covers an area of about 29 ha (71 acres), and about 2,700,000 m^3 (3,500,000 $yards^3$) of contaminated materials were placed in the cell yielding a total activity of 101 TBq (2,738 Ci) of radium-226 (US DOE, 2018a). The basic design of the disposal cell is similar to that at Grand Junction. An excavation was made into a hillside, and then the contaminated materials were covered with multiple barriers: compacted clay-sized material that is intended to be a barrier to radon and precipitation, coarse sand to facilitate drainage, then a fine-sand bedding layer to help secure an outer riprap layer that is used to protect the cell from erosion and volunteer plants. This disposal cell was designed to contain the radioactive wastes for at least 200 years (US DOE, 2018a). However, the Office of Legacy Management will be responsible for monitoring the Rifle Disposal Cell indefinitely.

The milling operations resulted in shallow groundwater contamination at both the Old Rifle Site (selenium, uranium, and vanadium) and the New Rifle Site (arsenic, molybdenum, nitrate, selenium, uranium, and vanadium). Similar to Grand Junction, US DOE selected natural flushing for the remediation of contaminated groundwater at the Rifle sites, once the source of the contamination was removed to the disposal cell. The reason that this case study was included in this chapter was that US DOE (2011b) questioned the efficacy of natural flushing. After a 14-year monitoring period, the concentrations of dissolved uranium in monitoring-well samples tended to remain constant or increase. As observed in Fig. 9.4, the concentrations of uranium in monitoring wells have both decreased and increased since 2011. Groundwater modeling had projected

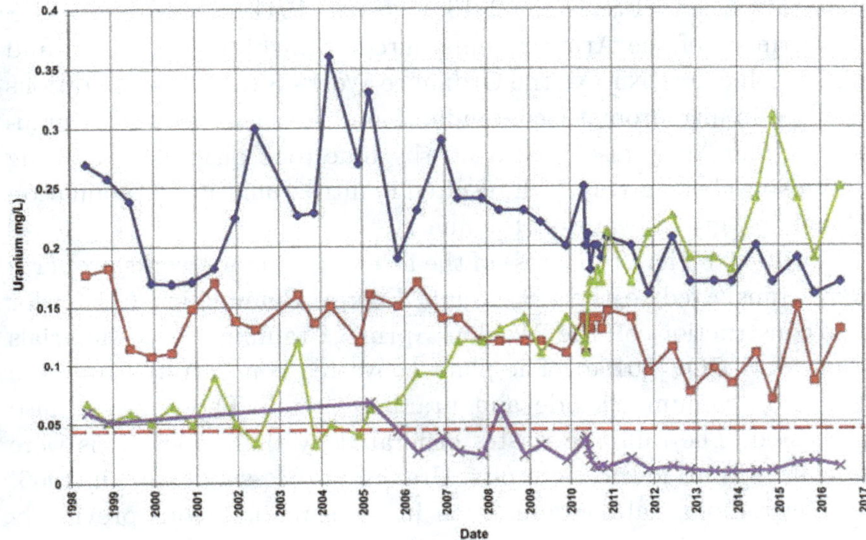

Figure 9.4. Concentrations of uranium in four monitoring wells at the Old Rifle site (US DOE, 2017).

natural flushing would reduce contaminant levels to background levels within 100 years. US DOE (2011b) concluded that the efficacy of natural flushing of mill-related uranium could not be reliably predicted at that time. Previous estimates did not take into account three potential sources of uranium:

1. Diffusion of uranium from low-permeable soils and sediments.
2. Mobilization of uranium from the water-unsaturated zone.
3. Movement of naturally occurring uranium in off-site groundwater into the Rifle sites.

9.4 The Weldon Spring (Missouri) Site Remedial Action Project

In 1941, the US government seized 6,974 ha (17,232 acres) of land southwest of the city of Weldon Spring, Missouri — in anticipation of World War II. Three small towns were evacuated, and 576 people were displaced (US DOE 2019c). From 1941 to 1945, the US Department of the Army manufactured trinitrotoluene (TNT) and dinitrotoluene (DNT) at the Ordnance Works Site. These operations resulted in nitroaromatic contamination of soils, surface and groundwater. The Army also used a nearby limestone quarry for burning and disposal of wastes which resulted in nitroaromatic contamination of soil and groundwater at the quarry.

In 1957, 88 ha (217 acres) of the former ordinance works property were transferred to the US Atomic Energy Commission (AEC) for the construction of the Weldon Spring Uranium Feed Materials Plant (US DOE 2019c). The plant converted concentrated uranium ore into uranium trioxide and uranium metal. Thorium was also processed. The raffinate wastes generated by these operations were discharged into four on-site pits. Uranium processing ended in 1966, leaving radiological contamination in the same locations previously contaminated by the Army. The site was included in this chapter because it is an example of a site cleanup where the contamination was a mixture of radioactive and chemical hazards (mixed wastes as defined in Chapter 4).

The buildings, raffinate pits, and the quarry were dormant for almost two decades. In 1984, however, US DOE became responsible for site cleanup. The Weldon Spring Remedial Action Project began in 1986. By 1990, the quarry, chemical plant area, and the raffinate pits were placed on the National Priorities List by the US EPA. Initial site assessments revealed the presence of a wide variety of radioactive and chemical contaminants. For example, the limestone quarry was about 300 m long and covered an area of about 3.5 ha (8.6 acres). In addition to the TNT and DNT process wastes disposed by the Army, the AEC disposed of uranium- and thorium-contaminated rubble and residues. Also, about 73,000 m^3 (2,578,000 $feet^3$) of bulk wastes containing arsenic, lead, nickel, selenium, and polychlorinated biphenyls were placed in the quarry (Haroun *et al.*, 1990). Polyaromatic hydrocarbons in the bulk wastes included volatile organic solvents such as acetone and methylene chloride. The mean concentration of 2,4,6-TNT in the surface soil within the quarry was 13 g/kg (Haroun *et al.*, 1990).

The technology chosen for the final disposition of the radiologically and chemical contaminated materials at the Weldon Spring site and the quarry was the construction of a disposal cell located on the Chemical Plant property. The removal of bulk wastes from the quarry was completed in 1995, and DOE completed the construction of the cell cover in 2001 (US DOE, 2019d). The five-sided cell covers an area of 18.2 ha (45 acres) and has 1:4 side slopes and is 28 m (91 feet) above grade at the center (Fig. 9.5). The design of the cell cover is similar to those at Rifle and Grand Junction. The top layer consists of limestone riprap over a sequence of aggregate bedding and drainage layers. Beneath these layers is a high-density polyethylene liner which is above a layer of compacted soil.

Because the disposal cell is located in a humid climate, it is a source of leachate which must be managed. It is assumed that the leachate is largely the result of the slow dewatering of the contaminated soils and sediments within the cell. Leachate samples are collected from the disposal cell and chemically characterized. The amount of leachate collected decreased from about 303 L/day (80 gallons/day) in 2011 to about 216 L/day (57 gallons/day) in 2018 (US DOE, 2019e). The mean uranium concentration in the leachate in 2018 was 0.86 Bq/L (23.2 pCi/L). Since 2010, the concentrations of uranium increased from about 0.59 Bq/L to 0.91 Bq/L (16 pCi/L to 24.5 pCi/L) at the end of 2019 (Fig. 9.6). The reason for the increase in uranium is unknown. Additional monitoring data are needed to fully understand the long-term chemical behavior of the disposal cell

Figure 9.5. The Weldon Spring disposal cell (courtesy of Eric Thomas, University of Denver).

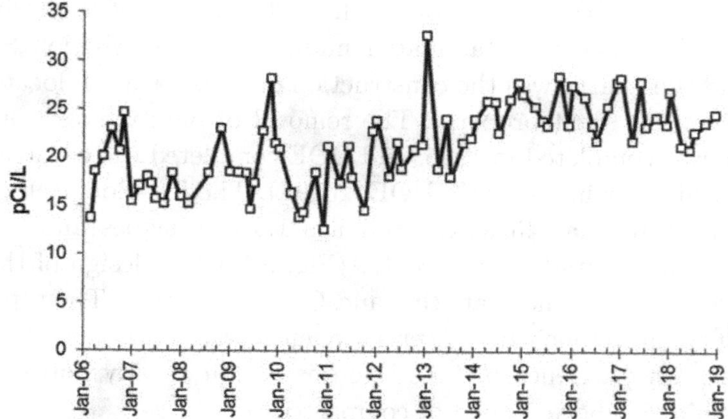

Figure 9.6. Concentrations of uranium in leachate samples collected from the Weldon Spring disposal cell from 2006 to 2019 (US DOE, 2019e).

leachate. The leachate is collected on-site and treated by passing it through filters and ion exchange media that is selective for uranium. The treated leachate is then sent to the Metropolitan St. Louis Sewer District for disposal.

The shallow groundwater is still contaminated at the former Chemical Plant and the quarry (US DOE, 2019e). The Chemical Plant monitoring network consists of collecting water samples at 48 monitoring wells, 4 springs and 1 surface-water location. The technology chosen for groundwater remediation is Monitored Natural Attenuation (MNA) coupled with restricting the use of groundwater during the period of remediation. Similar to natural flushing, MNA is a technique used to monitor the progress of natural attenuation processes that can reduce the concentration of contaminants in soil and groundwater once the source of the contaminants has been removed or isolated.

The raffinate pits were the primary source of uranium contamination in the shallow groundwater. Recent monitoring well data (US DOE, 2019d) suggest that in general, uranium concentrations on-site are constant or decreasing as a result of source removal, groundwater dilution, and dispersion. However, three monitoring wells that are near or within the areas for the former raffinate pits still contain relatively large concentrations (110–360 pCi/L) of uranium and have been increasing since 1999. Another example is trichloroethene (TCE). It is thought that the source of the TCE was the former Raffinate Pit 4. Three monitoring wells near the former pit have documented TCE concentrations of 160 to 34 μg/L in shallow groundwater samples in 2018 (US DOE, 2019e). Overall, concentrations have been slowly decreasing with time. TCE can be biodegraded in groundwater, but the process is generally slow (Roy, 2014). Under anaerobic conditions, TCE can be transformed into less chlorinated compounds in a sequence of TCE $\rightarrow$ cis-1, 2, dichloroethene (DCE) $\rightarrow$ vinyl chloride. Trace levels ($<1\,\mu$g/L) of DCE have been detected in monitoring well samples, but vinyl chloride has not been detected. It is suspected that groundwater dilution and dispersion were the only significant processes to attenuate TCE at the site (US DOE, 2019e).

9.5 The Fernald Preserve (Ohio)

The Fernald Preserve is located on the site of the former Feeds Materials Production Center which was a uranium-processing facility that was about 29 km northwest of Cincinnati, Ohio. The Feeds Facility was operated by the US Atomic Energy Commission from 1951 to 1989 and produced about 227,000,000 kg of uranium-metal products for the Nuclear Weapons Complex (Powell *et al.*, 2011; US DOE, 2018b). The facility also recycled used uranium from the reactors at the Hanford Site. Center operations resulted in the contamination of soil, sediments, surface water and groundwater with radionuclides, heavy metals and organic compounds.

Site remediation began during the 1990s. The site was divided into Operable Units (OU), each having a plan for site cleanup. Specifically, these units were six waste pits and a burn pit (OU 1), solid waste landfill, coal ash piles, and a lime sludge pond (OU 2), the former production area which contained about 200 buildings (OU 3), four below-grade silos containing metal oxides and contaminated soils (OU 4), and groundwater (OU 5). By 2006, all remedial actions were completed with the exception of OU 5 (US DOE, 2019f). Common to all the other units was that the contaminated soils, sediments, and debris were excavated, and either disposed off-site at the Energy *Solutions* Clive Disposal Facility (see Chapter 4), Andrew, Texas (Waste Control Specialists) or on-site. The remediation of and ecological restoration of the site was completed in 2006 at a cost of US$4.4 billion. The cleanup process was one of the largest efforts in the US (US DOE, 2019f).

Of relevance to this chapter is the On-Site Disposal Facility (OSDF). The OSDF is an above-ground waste disposal facility constructed to permanently store low-level radioactive and mixed wastes generated by decommissioning and soil excavation. The OSDF was designed to store 2,260,000 m^3, and it intended to isolate the wastes from the environment for at least 200 years and as long as 1,000 years to the extent possible (Kumthekar and Chiou, 2006). The OSDF consists of eight adjacent disposal cells and covers an area of 36 ha (89 acres) (Fig. 9.7). Construction of the OSDF began in 1997 and was completed in 2006. The eight cells were filled in sequence during site remediation. It was estimated that about 85% by volume of the waste placed in the OSDF was contaminated soils, and that 15% was decommissioning debris such as broken concrete and scrap metal (Kumthekar and Chiou, 2006; Hooten *et al.*, 2016).

Because the OSDF is located in a humid climate, the production of leachate from each disposal cell is minimized by being covered with a multilayer cover system (Fig. 9.8) to reduce infiltration into the cells. The cover is composed of topsoil over geomembranes which cover a layer of compacted clay that is placed over the wastes. Below the OSDF is a leachate collection system that is above geomembranes. At the bottom is a compacted clay liner — all to minimize the movement of leachate into groundwater. Leachate samples collected from the leachate collection system at each of the eight cells contained uranium concentrations between 3.39 μg/L and 686 μg/L in 2018 (US DOE, 2019f). Wastes were placed into the oldest cell (Cell 1) in 1997, and the concentrations of dissolved uranium increased in 2018. There is no evidence that leachate has leaked into the aquifer below the site.

Figure 9.7. Aerial view of the Fernald site in 2010 with the OSDF outlined in blue (Powell *et al.*, 2011). Used with permission of WM SYMPOSIA (Arizona, US).

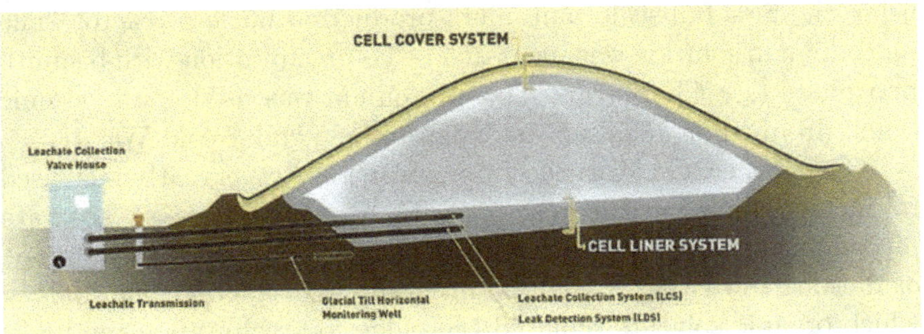

Figure 9.8. Cross-section of the on-site disposal facility at the Fernald Preserve (Hooten *et al.*, 2016). Used with permission of WM SYMPOSIA (Arizona, US).

The technology chosen to remediate the groundwater was source removal coupled with extracting and treating the groundwater in an above-ground facility (pump-and-treat). The treated water was periodically re-injected into the aquifer. The final remediation level is 30 μg/L. Since 1993, 6,437 kg of uranium have been extracted from the Great Miami Aquifer (US DOE, 2019f). The mass of uranium removed each year has been gradually decreasing, but a target date for completing the remediation of OU 5 has not been proposed. The groundwater below the Fernald Preserve in the Great Miami Aquifer, however, remains contaminated with uranium in isolated locations on-site.

9.6 The Hanford Site (Washington)

The Hanford Site, also known as Hanford Project, Hanford Works, Hanford Engineer Works and Hanford Nuclear Reservation, was established in a desert climate next to the Columbia River in Washington (US) by US Army Corps of Engineers in 1943 as part of the Manhattan Project. It originally covered an area of 1,740 km^2 (672 miles^2). At that time, the only purpose of Hanford was to produce plutonium by separating it from irradiated uranium slugs using the first full-scale plutonium-production nuclear reactor. The plutonium in solution was separated by co-precipitation with bismuth phosphate (see Chapter 5). This plutonium was used in the atomic bomb dropped on Nagasaki in Japan to help end World War II.

During the Cold War, the scope of Hanford was greatly increased to include nine nuclear reactors and five complexes to separate plutonium from spent uranium fuel using the REDOX (reduction–oxidation) and PUREX (plutonium uranium extraction) methods in which organic solvents replaced the older co-precipitation methods. Hanford produced about 54.5 tonnes (60.1 tons) of weapons-grade plutonium for the Nuclear Weapons Complex before the last reactor was shut down in 1987 (Gephart, 2010).

The operations and waste-management practices resulted in soil and groundwater contamination on a scale that dwarfs the case studies previously discussed in this chapter. In 1989, US DOE, US EPA, and the Washington State Department of Ecology entered into the Tri-Party Agreement to cleanup Hanford, and to better manage the radioactive, mixed, and chemical wastes that had accumulated. Hanford is the largest and most expensive environmental cleanup project in the US (Gephart, 2010). It is beyond the scope of this chapter to discuss Hanford in detail. There are many published studies and technical reports that have been written about the history, activities, environmental impacts, and the cleanup efforts at Hanford. Gephard (2010) provided an excellent summary of how wastes were managed at Hanford and is cited here at length. Hanford was included in the chapter as an example where complete environmental restoration may not be technically or economically possible.

The magnitude of environmental degradation at Hanford was exacerbated by the mismanagement of solid and liquid wastes. The separation of plutonium from the irradiated uranium slugs created an acidic raffinate at the processing plants which contained a variety of fission products. The "technology" chosen to manage this waste was to pump it into a total of 177 underground tanks. Because of the corrosive nature of the raffinate, sodium hydroxide was added to the liquid to neutralize the acids which resulted in chemical precipitates. While in the tanks, the precipitates settled to the bottom forming relatively insoluble sludges. It has been estimated that the Hanford tanks currently hold about 7.03×10^6 TBq (1.9×10^8 Ci) of radioactivity, and 170,000 tonnes (187,000 tons) of process chemicals (Gephart, 2010). Goel *et al.* (2019) estimated that the volume of mixed wastes in the tanks was about 212,000,000 L (56,000,000 gallons).

Gephart and Lundgren (1998) generalized that the tanks contain three phases: a vapor phase containing hydrogen, nitrous oxides, and ammonia over a liquid phase which is above solid salt cakes or sludges which are dominated by oxides, nitrates, and hydroxides of aluminum (Goel *et al.*, 2019). Deutsch *et al.* (2011) provided compositional data of sludge samples collected from single-shell tanks at the Hanford Site (Table 9.2). The investigators summarized that

Table 9.2. Estimated range in chemical composition of dried sludge samples collected from tanks C-103, 106, 202, 203, and S-112 at the Hanford Site (Deutsch et al., 2011).

Element	Range in concentration (dry weight)
Aluminum	1.36–27.1%
Calcium	0.01–4.65%
Iron	0.23–12.2%
Sodium	0.78–9.58%
Silicon	0.12–2.50%
Americium-241	0.01–2.05 mg/kg
Neptunium-237	1.30–9.0 mg/kg
Plutonium-237	8.02–435 mg/kg
Technitium-99	0.23–1.14 mg/kg
Uranium-238	0.002–50.5%

the chemical composition of the tank solids was extremely variable and that they reflected changes in the type of wastes disposed and post-placement chemical reactions. The reader should note that the mean uranium-238 content of sludge samples from tank C-203 was about 50.5%. Could this solid residue be a potential uranium resource?

Unfortunately, about 67 single-shell tanks at Hanford have leaked or are suspected of leaking into the soils below the tanks (Gephart, 2010). Under the conditions of the Tri-Party Agreement, the wastes in the tanks are to be retrieved, and sent to a Waste Treatment and Immobilization Plant (WTIP) which will use vitrification to immobilize the tank wastes. The WTIP is currently being designed and constructed, and a portion of the WTIP may begin operating in 2023.

The WTIP will cover 26.3 ha (65 acres) with four facilities: Pretreatment (to separate the low-activity, liquid wastes from the high-level tank solids), Low-Activity Waste Vitrification, High-Level Vitrification, and an Analytical Laboratory. The vitrified low-activity waste will be poured into stainless steel containers and stored on-site at Hanford. The vitrified high-level waste will also be poured into stainless steel containers for eventual placement in a federal geological repository.

Progress in retrieving of the wastes from the tanks, however, has been slowed by challenges such as mixing and chemically characterizing the wastes in the tanks, waste-pumping pressures, and pump failures (US DOE, 2013). Because of the delays and budget revisions, the timetable for the sequence of events for tank cleanup is uncertain. Pumpable liquid wastes have been transferred from aging single-shell tanks into newer double-shell tanks.

Hanford has a long history of releasing radioactive and chemical wastes into ponds, trenches, cribs (see the Preface), and injection wells. Low-level radioactive and transuranic wastes were dumped into shallow trenches. In 1997, there were 75 waste burial areas (Gephart, 2010). Not surprisingly, there are multiple plumes of contaminated groundwater. Contaminants of concern include chromium, trichloroethylene, carbon tetrachloride, tritium, iodine-129, and technetium-99 and strontium-90. Pump-and-treat technologies are being used to remediate the contaminated groundwater. One experimental technology that is being tested at Hanford is Apatite Sequestration. On a field-scale, a calcium-citrate-phosphate solution was injected into the subsurface resulting in the precipitation of apatite ($Ca_5(PO_4)_3$ (F, Cl, OH)). The apatite is being tested as a barrier to the movement of strontium-99 in groundwater. The strontium is sorbed from solution by the apatite, preventing it from spreading off-site (Fritz *et al.*, 2011).

Gephart (2010) speculated that about 37,000 TBq (1,000,000 Ci) of radioactivity and between 100,000 to 300,000 tonnes (110,000–330,000 tons) of chemicals remain in the soil and groundwater at Hanford. It is not possible to predict when site cleanup at Hanford will be completed, nor is it possible to predict the efficacy of the cleanup efforts during the decades to come. It seems very likely that some of the impacted areas will not be released to the public for unrestricted use and will require monitoring indefinitely.

9.7 Review Questions

The technologies or approaches used for site remediation of the five case studies are summarized in Table 9.3, beginning with the most passive and ending with the most complex and expensive. The application of a technology depended on geographic location, site geology, climate, size of the area impacted, and the extent and type of contamination. The remediation of sites that were impacted by the Nuclear Weapons Complex is an on-going effort. The most recent information (US DOE, 2019a), indicates that about 52 sites are at various stages of site remediation, and are scheduled to be transferred to US DOE's Office of Legacy Management by 2070.

1. Why was almost every site in the Nuclear Weapons Complex contaminated with radioactive or hazardous materials at the end of the Cold War?
2. What do the disposal cells in the case studies in this chapter have in common? How do they differ?
3. What are the possible limitations of Monitored Natural Attenuation and natural flushing?
4. Rank the five case studies in terms of their negative impacts on the local environment.

Table 9.3. Summary of technologies used for site remediation of the 5 case studies.

	Technology
Passive	Natural flushing
	Monitored natural attenuation
	Waste removal, disposal off-site
	Waste removal, on-site disposal cell (dry climate)
	Waste removal, on-site disposal cell with leachate capture and treatment (humid climate)
	Waste removal coupled with groundwater remediation (pump and treat, subsurface barriers)
Active	Waste removal and immobilization such as vitrification and disposal of treated wastes in a geological repository

5. Describe how the tank wastes at Hanford will be treated by vitrification.
6. Go to US DOE (2019a). Site Management Guide. US Department of Energy, Legacy Management. Update 23 or the current version at https://www.energy.gov/lm/downloads/site-management-guide.

Select one site that is scheduled to be transferred to US DOE's Office of Legacy.

Answer the following questions:

What is the background or history of the site?

Who owns the site now?

What is its current status?

What type of contaminants are known or suspected to be present? Is there soil contamination? Groundwater contamination?

What approaches and technologies will be used to restore the site? If no information is available, what technologies do you think could be feasible?

How will the wastes be disposed? Will any contaminated soil or groundwater be left on-site?

Will the site be released for unrestricted use?

Bibliography

Deutsch, W. J., Cantrell, K. J., Krupka, K. M., Lindberg, M. L., and Serne, R. J. (2011). Hanford Tank Residual Waste — Contaminant Source Terms and Release Models. *Applied Geochemistry*, 26, pp. 1681–1693.

Fritz, B. G., Szecsody, J. E., Vermeul, V. R., Williams, M. D., and Fruchter, J. S. (2011). *100-NR-2 Apatite Treatability Test: An update on Barrier Performance*. Pacific Northwest National Laboratory (Report PNNL-20252).

Gephart, R. E. (2010). A Short History of Waste Management at the Hanford Site. *Physics and Chemistry of the Earth*, 35, pp. 298–306.

Gephart, R. E. and Lundgren, R. E. (1998). *Hanford Tank Clean Up: A Guide to Understanding the Technical Issues.* Pacific Northwest Laboratory, Richland, Washington (Report PNL-10773).

Goel, A., McCoy, J. S., Pokorny, R., and Kruger, A. A. (2019). Challenges with Vitrification Of Hanford High-Level Waste to Borosilicate Glass — An Overview. *Journal of Non-Crystalline Solids: X*, 4 (https://doi.org/10.1016/j.nocx.2019.100033)

Haroun, L. A., MacDonell, M. M., Peterson, J. M., and Fingleton, D. J. (1990). Multimedia Assessment of Health Risks for the Weldon Spring Remedial Action Project. U.S. Department of Energy Legacy Management (Report 90-77.5).

Hooten, G., Hertel, W. A., Glassmeyer, C., and Broberg, K. (2016). Lessons Learned Concerning the Onsite Disposal Facility at the Fernald Preserve, Harrison, Ohio. In *Waste Management 2016 Conference*, March 6–10, 2016, Phoenix, Arizona (US).

Kumthekar, U. A. and Chiou, J. D. (2006). Lessons Learned from the On-Site Disposal Facility at Fernald Closure Project. In *Waste Management Conference*, February 26–March 2, 2006, Tucson, Arizona (US).

NGA. (2012). Twenty years of the Federal Facility Compliance Act: Lessons learned about the cleanup of nuclear weapons waste. National Governors Association. Washington, DC. Available at: http://www.nga.org/cms/home.html [Accessed 16 February 2020].

Powell, J., Abitz, R. J., Broberg, K. A., Hertel, W. A., and Johnston, F. (2011). Status and Performance of the On-Site Disposal Facility Fernald Preserve, Cincinnati, Ohio. In *Waste Management 2011 Conference*, February 27–March 3, 2011, Phoenix, Arizona (US).

Roy, W. R. (2014). The Environmental Chemistry of Organic Solvents, in Wypych, G. (ed.) *Handbook of Solvents, 2nd edition*, ChemTec Pub., Toronto, Canada, pp. 361–386.

US DOE. (1997). *Long-Term Surveillance Plan for the Estes Gulch Disposal Site Near Rifle Colorado.* US Department of Energy (Report DOE/AL/62350-235).

US DOE. (1998). *Work Plan for Characterization Activities at the UMTRA Project New and Old Rifle Sites.* US Department of Energy (Report MAC–GWRFL1.8).

US DOE. (1999). *Environmental Assessment of Ground Water Compliance at the Grand Junction UMTRA Project Site (Climax Uranium Millsite).* US Department of Energy (Report DOE/EA–1312).

US DOE. (2011a). *Evaluation of the Groundwater Monitoring Program at the Grand Junction, Colorado, Site.* US Department of Energy Legacy Management (Report no. S07776).

US DOE. (2011b). *Review of the Natural Flushing Groundwater Remedy at the Old Rifle Legacy Management Site, Rifle, Colorado.* US Department of Energy Legacy Management (Report number LMS/RFO/S07263).

US DOE. (2013). *Hanford Tank Waste Retrieval, Treatment and Disposal Framework.* US Department of Energy.

US DOE. (2016). *December 2015. Groundwater and Surface Water Sampling at the Grand Junction, Colorado, Site.* US Department of Energy Legacy Management (Report number LMS/GJO/SO1215).

US DOE. (2017). *October 2016. Groundwater and Surface Water Sampling at the Old and New Rifle, Colorado, Processing Sites.* US Department of Energy Legacy Management (Report number LMS/RFN/RFO/SO1016).

US DOE. (2018a). *Rifle, Colorado, Processing Sites and Disposal Site.* Fact Sheet. US Department of Energy Legacy Management.

US DOE. (2018b). *Fernald Preserve, Ohio.* Fact Sheet. U.S. Department of Energy Legacy Management.

US DOE. (2019a). *Site Management Guide.* US Department of Energy, Legacy Management. Update 23, (*Note that this guide is updated annually).*

US DOE. (2019b). *Grand Junction, Colorado, Site.* Fact Sheet. US Department of Energy, Legacy Management.

US DOE. (2019c). *Grand Junction, Colorado, Processing Site and Disposal Site.* Fact Sheet. US Department of Energy Legacy Management.

US DOE. (2019d). *Weldon Spring, Missouri.* Fact Sheet. US Department of Energy Legacy Management.

US DOE (2019e). *Weldon Spring, Missouri, Site Annual Report for Calendar Year 2018.* US Department of Energy Legacy Management. (Report number LMS/WEL/S23683).

US DOE. (2019f). *Fernald Preserve 2018 Site Environmental Report.* US Department of Energy Legacy Management (Report number LMS/FER/S23329).

Chapter 10

International Management of Radioactive Wastes

"Ah–what a journey–what a marvelous and extraordinary journey"

— Henry Lawson in "Journey to the Center of the Earth" by Jules Verne

10.1 Kingdom of Belgium

10.1.1 *Status of Nuclear Energy*

Belgium currently has two nuclear power plants that house a total of seven pressurized-water nuclear reactors. These reactors generate about 51% of Belgium's electricity (WNA, 2019). The future of nuclear energy in Belgium is uncertain. After numerous decisions and revisions, the current government has decreed a phase-out policy to take effect in 2025. By that deadline, all seven reactors are to be shutdown. The rationale for the current policy seems to be the result of safety concerns about the two power plants. These concerns have also been raised by anti-nuclear political parties. Public support for nuclear energy remains as a potential deterrent to the scheduled shutdown. It follows that there are no current plans to build new nuclear power plants.

Belgium has no uranium resources. Belgium once chemically reprocessed used nuclear fuel at the former Eurochemic plant in Dessel. Mixed oxide (MOX) fuel was formerly fabricated and used in Belgium. Reprocessing is no longer a priority in Belgium, and the Eurochemic plant and the MOX fuel fabrication plant are now being decommissioned (WNA, 2019).

10.1.2 *Waste Management*

About 70% of the radioactive wastes in Belgium are from the nuclear utilities (OECD-NEA, 2018). The wastes are managed by the National Agency for Radioactive Waste and Enriched Fissile Materials. The radioactive waste categories used in Belgium are similar to those outlined by IAEA (Appendix B). The management of the wastes is based, in part, on how they are classified:

Category A: Low- and medium-level, short-lived waste
The type of waste is dominated by beta and gamma sources with half-lives that are less than 30 years. An example is the waste created from routine power-plant operations.
Category B: Low- and medium-level, long-lived waste
This type of waste is dominated by alpha sources with half-lives longer than 30 years. Beta and gamma sources may also be present. An example is the waste materials created by the decommissioning of nuclear facilities.
Category C: Long-lived high-level waste
This type of waste contains alpha, beta, and gamma sources with both short and long half-lives, and produces "considerable" decay heat. An example of this category is the waste derived from chemically reprocessing spent nuclear fuel.

All radioactive wastes are currently stored in interim, above-ground buildings at the Belgoprocess facility. Low-level wastes are stored in Buildings 150 and 151. Medium-level wastes are placed in Buildings 127, 155, and 270. High-level and "very high-level" (OECD-NEA, 2018) wastes are placed in vertical, steel-lined wells in Buildings 129 and 136. These buildings are used to stored vitrified high-level wastes resulting from reprocessing at the former Eurochemic plant, and from reprocessing Belgium spent fuel at

Figure 10.1. Building 136 for storing vitrified high-level waste (Huys *et al.*, 2016). Used with permission of Belgoprocess.

La Hague in France. Building 136 (Fig. 10.1) was constructed to survive earthquakes and terrorist attacks (OECD-NEA, 2018). Spent fuel is currently stored by wet storage at the Tihange Power Plant. At the Doel Power Plant, spent fuel is stored in Transnuclear TN24 dry dual-purpose casks (see Chapter 5) that are placed in an above-ground building near the plant.

There is a national effort in Belgium to site and build a geological repository for the long-term management of Category B and C wastes. Beginning in about 1974, field research by the Belgium Geological Survey and the National Nuclear Research Center began with the construction of an underground research laboratory. This project is called the High Activity Disposal Experiment Site (HADES) and was constructed at a depth of about 225 m in the Boom Clay (Fig. 10.2).

The Boom Clay is an Oligocene-age marine clay and was deposited as a seafloor sediment (Vandenberghe *et al.*, 2014). The suitability of the clay as a geological repository is currently under study. Research includes excavation techniques, anaerobic corrosion of steel containers, geochemical processes in the clay, geochemical behavior of the radionuclides once released from the waste, and the migration of radionuclides and gases in the clay (Neerdael and Vokaert, 2001; Aertsens *et al.*, 2013). A timeline for site selection and the construction of the repository has not been proposed (OECD-NEA, 2018). HADES will not be converted into the repository.

10.2 Federative Republic of Brazil

10.2.1 *Status of Nuclear Energy*

Brazil has one nuclear power plant that contains two pressurized-water reactors. Nuclear power generates about 3% of Brazil's electricity (WNA, 2019). A third reactor is partially constructed, but the project has been delayed because of economic difficulties. There are four research reactors in operation including a TRIGA reactor (see Chapter 7). Brazil has about 278,000 tonnes of uranium which is about 5% of the world's resources (WNA, 2019). Brazil has two uranium mines and mills (Filho *et al.*, 2015). It also has the capacity for uranium conversion, fuel fabrication, and enrichment, but has relied on the cooperation of France and Germany for much of the production of civilian reactor fuel.

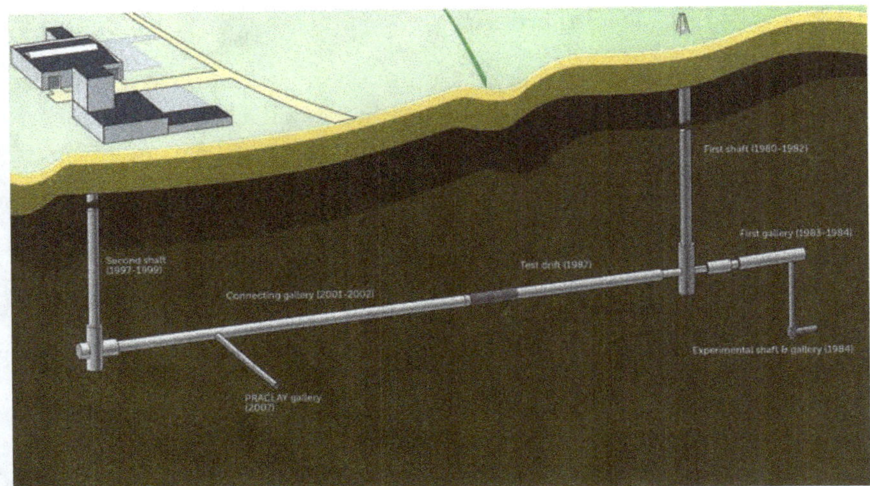

Figure 10.2. The HADES underground research laboratory (from Belgium Nuclear Research Center, 2020). Available at: http://science.sckcen.be/en/Facilities/HADES [Accessed 16 February 2020]. Used with permission of EVS EURIDICE GIE.

10.2.2 *Waste Management*

The National Nuclear Energy Commission is responsible for the management of all radioactive wastes in Brazil. Brazil uses the criteria given by IAEA for radioactive waste classification given in Appendix B (Filho *et al.*, 2016). Low- and intermediate-level wastes are derived from the operation of the two power reactors, uranium mining and milling, radioisotope production, research, and environmental restoration (Filho *et al.*, 2016). The low-level wastes specially generated by the Angra Nuclear Power Plant are stored in an on-site, above-ground building (Fig. 10.3). Spent nuclear fuel is currently managed by wet storage (see Chapter 5). Spent fuel is currently not considered as a high-level waste because the Brazilian government has not made a decision about the long-term disposition of the material. Chemical reprocessing and placement into a geological repository are both being considered (Filho *et al.*, 2016). A near-surface disposal facility for low-level wastes has been proposed, but the site-selection process has not begun.

10.3 Canada

10.3.1 *Status of Nuclear Energy*

Canada has 19 Candu (Canada deuterium uranium) reactors. Eight are located at the Bruce Nuclear Generating Station, six are located at the Pickering Nuclear Generating Station, and four are at the Darlington Nuclear Generating Station — all in Ontario. One is located at the Point Lepreau Nuclear Generating Station in New Brunswick (WNA, 2019). Canada currently has five research reactors operating on university campuses.

Canada has about 514,000 tonnes of uranium which is about 8% of the resources of the world. In 2017, it produced 13,116 tonnes of uranium, making it the second largest producer in the world after Kazakhstan. As a consequence, Canada has uranium mines and mills, and can refine, convert UO_3 to uranium dioxide (UO_2), and fabricate fuel bundles for domestic use and for exportation. Canada does not chemically reprocess spent nuclear fuel.

Figure 10.3. Storage of low-level radioactive wastes at the Angra Power Plant (Filho *et al.*, 2015). Used with permission of the Brazilian Nuclear Energy Commission.

10.3.2 *Waste Management*

Canada's radioactive waste classification categories are adapted from the IAEA (Appendix B). These categories are low-level, intermediate-level, and high-level wastes. The Nuclear Waste Management Organization (NWMO) and the Canadian Nuclear Laboratories (CNL) are responsible for managing spent nuclear fuel. The nuclear utilities and CNL are responsible for managing all low- and intermediate-level wastes which are currently stored in an above-ground concrete warehouse at the Western Waste Management Facility (WNA, 2019). Ontario Power Generation proposed to construct a deep, geological repository for all low- and intermediate-level wastes 680 m below the Western Waste Management Facility in an Ordovician-age limestone (see Chapter 6).

Spent nuclear fuel is initially placed in wet storage at the source power plants for 7–10 years to allow for a reduction in decay heat and radiation. After this initial cooling period, the spent fuel is transferred to dry storage. It is either placed in concrete canisters, dry-storage containers, or MACSTOR units, all of which were developed by Atomic Energy of Canada Limited. MACSTOR or Modular Air-Cooled Canister Storage is a horizontal concrete vault. Above-ground storage casks are discussed in Chapter 5. NWMO is conducting a site-selection process for a geological repository for spent fuel. While the engineering design of the repository is evolving, Jensen *et al.* (2016) reported that spent CANDU fuel will be placed in steel containers that are coated with a 3-mm thick layer of copper (Fig. 10.4). The canisters will be inserted into a bentonite-buffer box within the repository. NWMO expects to have the repository available by 2035 (WNA, 2019).

10.4 The People's Republic of China

10.4.1 *Status of Nuclear Energy*

China currently has 45 commercial nuclear reactors located among 12 power plants. All of the power plants are located along the eastern part of the country. China does not operate a single style of reactor

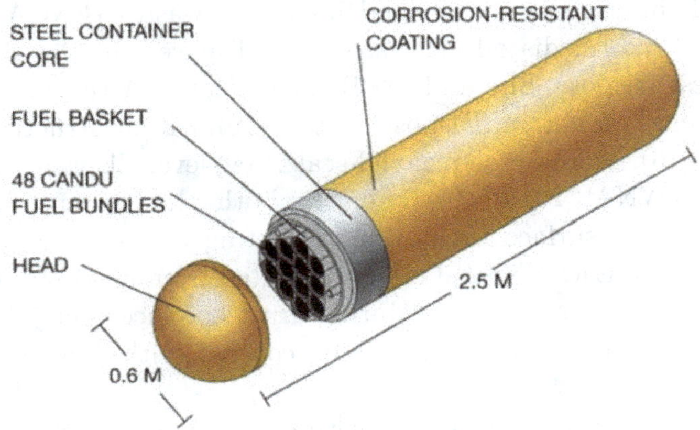

Figure 10.4. Prototype of Canadian repository container. Used with permission of the Nuclear Waste Management Organization.

but makes use of a variety of designs including French and Chinese pressurized-water reactors, one CANDU reactor, two water-water energetic reactors, and three self-styled advanced Chinese pressurized water reactors (WNA, 2019). China has 19 research reactors. China has an ambitious plan to increase its nuclear energy capacity: about 56 new reactors are either under construction or are planned. About 170 new reactors have been proposed. The major motivation for China's energy goals is the poor air quality in urban areas resulting from the use of fossil fuels. About 73% of China's electricity was derived from coal in 2015 (WNA, 2019). China has about 5% of the world's uranium resources (290,400 tonnes) and is currently operating seven uranium mines. China has the technology to convert and enrich uranium, and to fabricate uranium fuel. However, it currently relies on international sources for much of its uranium. Securing enough uranium to support China's ambitious plans may be a national challenge.

10.4.2 *Waste Management*

Radioactive waste management in China can be described as tenuous and evolving. Radioactive waste categories in China are low-level, intermediate-level, and high-level wastes (FNCA, 2007). Low- and intermediate-level wastes are disposed at two sites: the Northwest Disposal Site and the Beilong Center. A third site called the Feifeng Mountain Disposal Site is under construction (Shu *et al.*, 2016). Additional regional-scale, low-level disposal sites are planned (WNA, 2019). Common to both the current facilities, each is a near-surface excavation consisting of concrete vaults in disposal cells. Each disposal cell is protected from precipitation by a mobile, rain exclusion shelter during drum emplacement (US DOE, 2011). At closure, each unit will be covered with 2–5 m of local overburden.

At present, nearly all spent nuclear fuel in China is placed in wet storage at the source power plant. It appears that the current wet-storage capacity will be adequate until 2025–2043, depending on the power plant (Shu *et al.*, 2016). The storage pools at one power plant (Daya Bay) however, are full. China National Nuclear Corporation (CNNC) Everclean is planning to acquire Holtech International HI-STAR 100 MB storage casks (see Chapter 5) (WNA, 2019).

It is the policy of the Chinese government that the spent fuel from light-water reactors should be reprocessed, and that the vitrified waste products from chemical reprocessing be disposed in a centralized geological repository (Shu *et al.*, 2016; Wang *et al.*, 2016). A pilot-scale reprocessing plant using the Purex process (see Chapter 5), has been in operation since 2015 (Wang *et al.*, 2016). CNNC, however, has no experience with mixed oxide (MOX) fuel fabrication and use. Some type of MOX facility may be available after 2020 (WNA, 2019).

A general location for a geological repository has been selected, but a specific location has not been chosen (Wang *et al.*, 2016). It is envisioned that the Precambrian-age Beishan Granite will be the host material for a geological repository in northern China. It has been proposed that an underground research laboratory will be constructed and used for *in situ* investigations until about 2040. It has also been proposed that the repository will be available for the disposal of vitrified wastes by about 2050. Preliminary designs suggest that the vitrified wastes will be placed in a carbon–steel

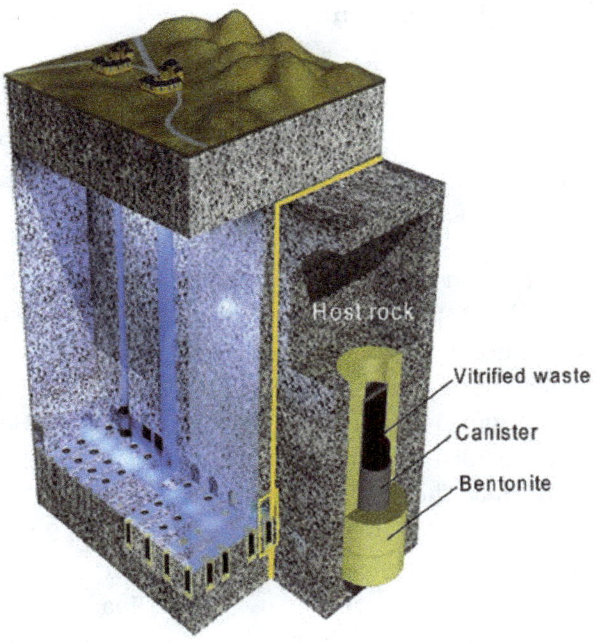

Figure 10.5. Conceptual model (not to scale) for China's High-Level Waste repository in the Beishan Granite (Wang *et al.*, 2016). Used with permission of ISRM Commission on Radioactive Waste Disposal.

canister which will inserted into a bentonite buffer in the granite at a depth of 500 m (Fig. 10.5).

10.5 Czech Republic

10.5.1 *Status of Nuclear Energy*

The Czech Republic has six water–water energetic reactors that are housed within two power plants. These reactors produce about 29% of the country's electricity (WNA, 2019). The government supports nuclear energy as an approach to reduce carbon dioxide emissions. Two new power plants have been proposed. The Czech Republic once mined uranium, but the depletion of uranium resources resulted in the closure of the last mine in 2017. The country now relies on Russia for fuel fabrication.

10.5.2 *Waste Management*

The Radioactive Waste Repository Authority (RAWRA) is responsible for the management of all radioactive waste in the Czech Republic. Radioactive waste classification follows that of the IAEA (Appendix B). The Czech Republic has four repositories low- and intermediate-level wastes:

1. The Richard Repository. Originally an underground limestone mine, it has been used for the disposal of "institutional" low-level wastes since 1964 (Haverkamp *et al.*, 2005). The repository is an extensive network of tunnels and caverns within a limestone layer that is about 4 m thick (Fig. 10.6). The institutional wastes are from the health sector, industrial sources, and research. It is estimated that the Richard Repository will be available for additional wastes until about 2070 (OECD-NEA, 2018).
2. The Hostim Repository is another former underground limestone mine. It was also used to dispose of radioactive institutional wastes. It was closed in 1965 and has since been backfilled with concrete and sealed.
3. The Bratrstvi Repository is a former uranium mine that is used for the disposal of institutional wastes. In 2012, one of the disposal tunnels was sealed (OECD-NEA, 2018). The repository is still in use.
4. The Dukovany Repository is the largest facility of the four and was built specifically for low- and intermediate-level wastes generated by the two nuclear power plants. It is an above-ground collection of 112 concrete vaults (OECD-NEA, 2018). The repository has a capacity of $55,000\,\text{m}^3$ or about 130,000 200-L steel drums. Prior to placement in the vaults, the wastes in the drums are mixed with either bitumen or concrete to help prevent the wastes from leaching. The repository has been in operation since 1995. When all the vaults are filled, a final cover will consist of layers of concrete and soil.

Spent nuclear fuel is currently managed by wet and dry storage at the source power plants. Dry storage casks have been used since

Figure 10.6. The Dukovany ISFSF (CRNR, 2005). Used with permission of NPP Dukovany.

1995 (see Chapter 5). The Dukovany Independent Spent Fuel Storage Facility (ISFSF) is situated inside the Dukovany Nuclear Power Plant and is designed for dry storage of spent fuel in CASTOR-440/84 casks (Fig. 10.6). Chemical reprocessing of spent fuel is not regarded as economical by the RAWRA. A geological repository for the spent fuel is planned for the Czech Republic (Slovak and Woller, 2016), but details are not yet available, and the site-selection process is ongoing. The proposed disposal depth is 500 m in the Bohemian Massif (a Pre-Cambrian granite). It is envisioned that the spent fuel will be placed into stainless-steel containers that will be imbedded within some type of bentonite seal (Slovak and Woller, 2016). The proposed date to have a functional repository in about 2065.

10.6 Republic of Finland

10.6.1 *Status of Nuclear Energy*

Finland has two water–water energetic reactors at the Loviisa Nuclear Power Plant and two boiling water reactors at the Olkiluoto Nuclear Power Plant (WNA, 2019). The four reactors provided about

34% of the country's electricity in 2016. The Finnish government supports nuclear energy as a viable way to reduce coal use and the country's dependence on hydroelectricity. A fifth reactor (a European Pressurized Water Reactor) is currently under construction and may be operational by 2020 at the Olkiluoto Plant. A sixth reactor and a new power plant have been proposed by Fennovoima Oy, a consortium of industrial and energy companies. Finland has few mineral resources. Coal must be imported from Russia and Poland (WNA, 2019). There are no uranium mines in Finland. The country imports uranium from Canada, Australia and Africa. It depends on Canada, France, Russia, Sweden and Spain for fuel enrichment and fabrication.

10.6.2 *Waste Management*

In Finland, the electric utilities are responsible for radioactive waste management. The classification system used in Finland for radioactive wastes resembles the IAEA system (see Appendix B). Low- and Intermediate-Level Waste (LILW) are from the operations of the two power plants, Small User Waste (low-level waste from medical, research, and industrial applications), and Spent Nuclear Fuel (OECD-NEA, 2018).

Finland is the world leader in constructing geological repositories for radioactive wastes. The LILW wastes generated by the Okiluoto Plant are placed in the Okiluoto VLJ Repository which is a geological repository. The abbreviation VLJ was derived from the Finnish expression "voimalaitosjdte" which means "reactor operating waste" (Bergström *et al.*, 2011). LILW from the Loviisa Plant are disposed at the Loviisa VLJ Repository which is also a near-surface geological repository. Both of these LILW repositories are described in detail in Chapter 6.

The final disposition of spent nuclear fuel in Finland is managed by Posiva Oy. A deep, geological repository is planned for the disposal of all spent fuel in Finland. To arrive at that goal, an underground research laboratory called ONKALO (Finnish for small cave) was constructed to depth of 450 m. ONKALO will become part of the larger Okiluoto Repository which may be operational in 2023 (see Chapter 6). The Okiluoto Repository may become the world's first geological repository for spent nuclear fuel.

In 2016, Fennovoima Oy proposed to construct their own geological repository for the spent fuel created by the proposed nuclear plant (Fennovoima, 2016). The site selection is to be made in the 2040s. It is interesting that this proposal could lead to the unprecedented situation of having *two* deep geological repositories for spent nuclear fuel in the *same* country.

10.7 French Republic

10.7.1 *Status of Nuclear Energy*

France is a world leader in nuclear energy and radioactive waste management. France has 58 pressurized-water reactors that are housed within 19 power plants (WNA, 2019). Électricité de France operates all the nuclear power plants. One European Pressurized Reactor is under construction. France also has 11 research reactors (US DOE, 2011). France currently generates about 75% if its electricity from nuclear power. Both the public and the government strongly support nuclear energy, but the government has mandated that the contribution of nuclear energy for creating electricity be reduced to 50% by 2025. France has a closed fuel cycle; it has facilities for uranium conversion, enrichment, fuel fabrication, used fuel reprocessing and waste treatment, and mixed-oxide (MOX) fuel fabrication (WNA, 2019). About 17% of France's electricity is derived from MOX fuel.

10.7.2 *Waste Management*

The National Radioactive Waste Management Agency (Agence nationale pour la gestion des déchets radioactifs or Andra) is responsible for the management of all radioactive waste in France (OECD-NEA, 2018). The waste classification system used in France is somewhat similar to the outline provided by the IAEA (Annex B),

Table 10.1. Radioactive waste classification system used in France and characteristics (SNR, 2017).

Waste category	Waste characteristics
Very Short-Lived	Medical and research sources. Half-life less than 100 days.
Very Low-Level (VLL)	Decommissioning wastes and operational wastes. Radionuclides with a half-life that is ≤31 years. Specific activity less than 100 Bq/g.
Low-Level and Intermediate-Level, Short-Lived (LIL-SL)	Decommissioning wastes and operational wastes.
Low-Level, Long-Lived (LL-LL)	Graphite and radium-containing wastes. Specific activity 10–10^5 Bq/g. Radionuclides having a half-life longer than 31 years.
Intermediate Level, Long-Lived (ILW-LL)	Reprocessing wastes. Specific activity 10^6–10^9 Bq/g. Radionuclides having a half-life longer than 31 years.
High-Level Waste (HLW)	Vitrified wastes from reprocessing. Specific activity greater than 10^9 Bq/g.

but it has been expanded to reflect the activities of the radionuclides and their half-lives (Table 10.1).

Historically, the first near-surface disposal facility for Low- and Intermediate-Level Short-Lived waste is called the Centre de stockage de la Manche. It also known as the CSM facility, Centre de la Manche disposal facility or simply the Manche repository. It first accepted waste packages in 1969. The wastes were placed in concrete disposal cells (Dutzer *et al.*, 2006). About 527, 225 m^3 of waste were disposed in the 12-ha site before it closed in 1994. Between 1991 and 1997, Andra constructed a multilayer cover over the disposal cells. The disposal drums were buried with a succession of materials that are intended to reduce and divert infiltration, and to provide protection from animal and plant intrusions. The layers consisted of (from bottom to top) compacted coarse-grained shale and sandstone, a fine-sand drainage layer, a 5.5-mm thick bituminous-based geomembrane, a second fine-grained sand drainage layer, a second layer of coarse-grained shale and sandstone, and as the uppermost cover, 0.3 m of soil (Dutzer *et al.*, 2006). The reader may find it interesting to compare the Manche final cover with the design of the disposal cells used by the US DOE for uranium milling wastes (see Chapter 9). The structural integrity of the final cover at Manche is being monitored by Andra for the foreseeable future.

The Price of Success?

It may be puzzling to the reader why the French government has mandated a reduction of nuclear energy available for creating electricity. In 2015, the French government promulgated The Energy Transition for the Green Growth Act (MESDE, 2015). The Act declares that the use of renewable sources of energy is to increase from 15% in 2014 to 40% in 2025. A major motivation for the Act is France's desire for greater energy independence, but France does not have the domestic uranium resources needed to sustain its successful nuclear energy program. The country must depend on uranium from Canada, Niger, Australia, Kazakhstan, and Russia. This unavoidable dependency may be in part why France seeks to reduce its use of nuclear energy.

A repository for Low- and Intermediate-Level Short-Lived wastes was constructed by Andra to take the place of the Manche facility. Its replacement is called the Centre de Stockage de l'Aube (CSA). It is also known as Soulaines-Dhuys Repository, or simply the Soulaines repository. The 200-L waste drums are stacked in concrete vaults in the near-surface repository. The vaults are intended to isolate the waste drums from groundwater for 300 years which is about 10 half-lives of cesium-137. Once the disposal cell is filled, it is capped with concrete and covered with soil (US DOE, 2011).

A third repository is in operation for Very Low-Level wastes. It is called Centre Industiel de Regroupement, d'Entreposage et de Stockage or CIRES. It is also known as the Morvilliers repository. It is a near-surface, 45-ha facility (US DOE, 2011). It was designed to

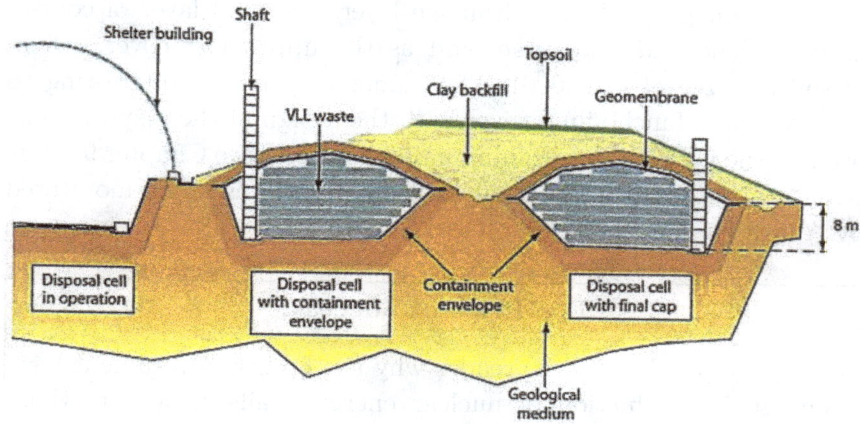

Figure 10.7. Cross-section of the CIRES Very Low-Level Waste repository (US DOE, 2011).

store $650,000\,m^3$ of waste, and the first waste shipment was emplaced in 2003. The wastes are placed in trenches excavated in soil. When the trench is full, it is filled with sand, then covered with compacted clay and a geomembrane. The resulting disposal cell is then covered with soil (Fig. 10.7).

Used nuclear fuel is initially placed in wet storage at the source power plant. After a period of time for the decay heat to cool, the used fuel is placed in transportation flasks and moved to pools at the La Hague plant where the fuel may be stored for about 5 years (Valery *et al.*, 2015). In France, used nuclear fuel is not classified as a waste, but it is regarded as a resource. It is the policy of the French Government to chemically reprocess used nuclear fuel to extract uranium and plutonium to fabricate mixed-oxide fuel (MOX). Reprocessing is discussed in Chapter 5. Commercial reprocessing using the Purex process is currently conducted at the La Hague plant. Sombret (1993) provided an excellent review of the history and details about the French Industrial Vitrification Process that is used to treat the raffinate waste created by reprocessing. An excellent review of spent fuel reprocessing in France was provided by Schneider and Marignac (2008). Ultimately the reprocessing waste is converted into a molten vitrified mass that is cooled in stainless steel containers. The vitrified solids are stored on an interim basis in below-surface wells inside a storage facility at the La Hague plant.

France plans to develop a geological repository for the vHLW. The research needed to build a repository began in 1999 with the construction of an underground research laboratory (URL) near the village of Bure. This underground laboratory is discussed in Chapter 6. Based on decades of research and field work, Andra proposed to build the repository about 5 km from the Bure URL (Farin *et al.*, 2016). The repository is named the Centre Industriel de Stockage Géologique (Cigéo) or Industrial Center for Geological Disposal. It is planned that Cigéo will be designed and operated for the reversible disposal of the vitrified wastes and long-lived intermediate-level wastes. The vitrified-waste containers will be placed in steel overpacks prior to placement into the Callovo-Oxfordian formation. Additional engineered barriers are regarded as unnecessary (Farin *et al.*, 2016). Andra plans to have Cigéo operational by 2025 and it will be designed to accommodate 10,000 m^3 of vHLW (WNA, 2019).

10.8 Federal Republic of Germany

10.8.1 *Status of Nuclear Energy*

Germany currently has seven operating nuclear power reactors (one boiling-water reactor and six pressurized-water reactors) located in seven power plants. These reactors currently create 12% of the country's electricity (WNA, 2019). The future of nuclear energy for power generation in Germany is bleak. In 2011, the German government decided to phase out nuclear energy for the production of electricity (Bräuer, 2016). All seven power plants are to be shut down from 2019 to 2022. The decision for the phase-out plan is driven by anti-nuclear politics. Public opinion is also opposed to nuclear energy. Germany has a long history of vacillating between supporting and opposing nuclear energy for future use. It seems likely that part of the anti-nuclear movement has been fueled by the government's poor record of managing radioactive wastes.

Germany once operated relatively small uranium mines, but the mines are now closed. All uranium is now imported from Canada, Australia, and Russia (WNA, 2019). Germany has the capacity for uranium enrichment and fuel fabrication. Germany chemically reprocessed used nuclear fuel using the Purex process until the governmental terminated this option in 1994.

10.8.2 *Waste Management*

The Federal Company for Radioactive Waste Disposal (Bundesgesellschaft für Endlagerung) is currently responsible for all radioactive waste management and will coordinate the management of disposal sites with the Federal Institute for Geosciences and Natural Resources (Bundesanstalt für Geowissenschaften und Rohstoffe). Radioactive waste classification in Germany does not use the IAEA categories given in Appendix B. Germany classifies radioactive wastes into two broad categories: Heat-Generating Waste (HGW) and Non-Heat Generating Waste (non-HGW) (US DOE, 2011). In general, non-HGW is comparable to low-level waste combined with low- and intermediate-level waste that does not generate a significant amount of decay heat. The HGW category includes spent nuclear fuel and reprocessing wastes (Bräuer, 2016). Germany does not differentiate waste type by the presence of short-lived or long-lives radionuclides. Very low-level wastes are equivalent to a clearance category in Germany and may by disposed in conventional landfills (US DOE, 2011).

Germany chose not to have near-surface disposal sites for radioactive wastes (US DOE, 2011). The approach used was to adapt former mines as repositories for the disposal of non-HGW. These repositories are the Asse II (a former potash and salt mine in a salt dome), the Morsleben Repository (a former salt mine), and the Konrad Repository (a former iron ore mine). All three repositories are discussed in Chapter 6. The Asse and Morsleben Repositories are now closed, and the Konrad Repository may be operational by 2027 (Bräuer, 2016). A salt dome near Gorleben was under investigation as a possible geological repository for both non-HGW and HGW, but the project has been suspended for political reasons. Germany currently has no definite plans for the disposal of HGW. Spent nuclear fuel is stored in dry-storage casks after cooling at the source power plant. Some type of geological repository may be selected by 2031, and commissioned by about 2050 (Bräuer, 2016).

10.9 Japan

10.9.1 *Status of Nuclear Energy*

Japan currently has nine pressurized-water reactors that are operating in five power plants (WNA, 2019). Seventeen additional reactors (pressurized- and boiling-water reactors) may be re-started pending safety approvals by the Nuclear Regulation Authority. The future of nuclear energy in Japan is uncertain since the Great East Japan Earthquake of 2011 further eroded the public's support for nuclear energy. Japan, however, has the capacity to enrich uranium and to reprocess used fuel by using the Purex process. Japan also has the ability to treat reprocessing wastes by vitrification. Japan does not have uranium resources, and must import uranium from Australia, Canada, and Kazakhstan. Based on imported uranium, Japan seeks to develop a complete nuclear fuel cycle (WNA, 2019).

10.9.2 *Waste Management*

Radioactive waste categories in Japan are not based on the IAEA system (Appendix B). Radioactive wastes are classified into two broad categories: High-Level Waste (HLW) and Low-Level Waste (LLW) (NWP, 2016). HLW are the wastes created by reprocessing such as the vitrified solids. Low-level wastes are divided in four categories that are based on their origins:

1. Power Plant Wastes (referred to as operational wastes in other countries).
2. Long-Lived, Low-Heat Generating Wastes (also informally referred to as Transuranic Wastes and as Mixed Oxide Fuel Fabrication Wastes).

3. Uranium Wastes (derived from uranium enrichment and fuel fabrication).
4. Facility Wastes (derived from research and medical facilities).

The Ministry of Economy, Trade, and Industry is responsible for radioactive waste management policy (US DOE, 2011). Japan Nuclear Fuel Limited operates the Rokkasho Low Level Disposal Centre. The Rokkasho facility has been in operation since 1992, and it is composed of near-surface disposal trenches that were excavated into the local bedrock (Fig. 10.8). Within the trenches are concrete vaults that are divided into cells in which 200-L metal drums are stacked horizontally (Bergström *et al.*, 2011). The majority of the low-level wastes sent to Rokkasho are transported from the source power plant after the waste-containing drums have been sealed with concrete (US DOE, 2011). When the vaults are filled, a concrete lid is added to each vault, and the trench is filled with a mixture of bentonite clay and soil, then buried with 4 m of soil. A new low-level facility is planned to be available in about 2023 (WNA, 2018).

Spent fuel is not classified as a waste in Japan. It is the policy of the Japanese government to chemically reprocess used nuclear fuel. This policy is promoted because Japan has few natural resources, and because there are economic benefits associated with using MOX

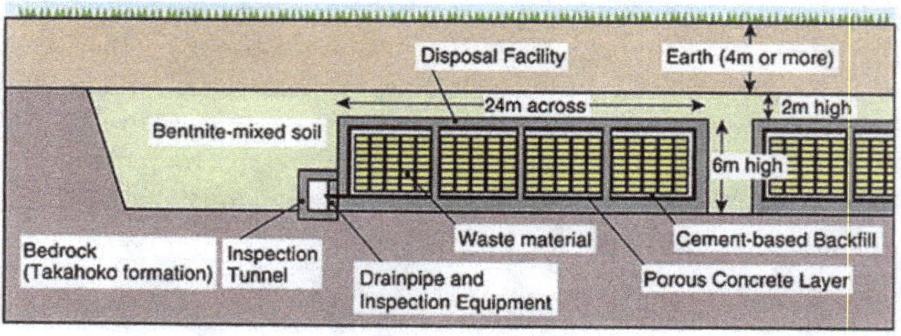

Figure 10.8. Rokkasho Low Level Disposal Centre (Japan Nuclear Fuel Limited, 2020). Available at: https://www.jnfl.co.jp/en/business/llw/ [Accessed 16 February 2020]. Used with permission of Japan Nuclear Fuel Limited.

fuel in the power plants. Japan Nuclear Fuel Limited currently stores used nuclear fuel for 15 years before reprocessing. Japan reprocessed fuel at the former Tokai Pilot Reprocessing Plant in addition to transporting used fuel to France and the UK for reprocessing. Japan Nuclear Fuel Limited plans to commission a new reprocessing plant at Rokkasho in 2021 (WNA, 2019). Meanwhile, used nuclear fuel is stored at the source power plants by wet and dry storage. Japan has been using dry storage casks since 1995 (see Chapter 5). The Recyclable Fuel Storage Company is constructing an above-ground storage facility for used nuclear fuel in dry storage containers at Mutsu. It may be operational by 2019. Japan Nuclear Fuel Limited operated a pilot-scale vitrification plant at Tokai, and canisters containing the resulting solid wastes are stored at the Vitrified Waste Storage Centre at Rokkaho.

Japan plans to construct a deep geological repository for vitrified high-level wastes and transuranic wastes. To accomplish this goal, the Nuclear Waste Management Organization (NUMO) was established for the site-selection process (Deguchi *et al.*, 2016). The Final Disposal Plan of 2008 mandated the construction of a repository with engineered and geological barriers. In order to conduct the field-scale research needed for the site-selection process, the Japan Atomic Energy Agency established two underground facilities: the Mizunami Underground Research Laboratory (ULR) and the Horonobe Underground Research Laboratory. Construction of the Mizunami URL began in 2002. It consists of two parallel, vertical shafts that currently extend to a depth of 500 m (Deguchi *et al.*, 2016). At the 500-m level, two tunnels were created into the Cretaceous-age Toki Granite (Fig. 10.9). Construction of the Horonobe URL began in 2005. It currently consists of three vertical shafts to a depth of 350 m in a Neogene-age mudstone in the Koetoi Formation (Deguchi *et al.*, 2016). Research at both underground laboratories includes hydrogeological, geochemical, geophysical, rock mechanical investigations, and geological mapping (Osawa *et al.*, 2015; Deguchi *et al.*, 2016). NUMO is currently seeking communities to volunteer for the site-selection process, but no communities have expressed an interest to date. NUMO, however, expects that the site-selection

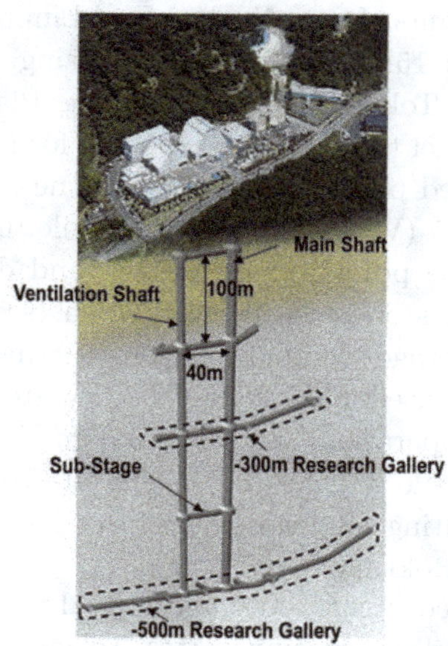

Figure 10.9. Mizunami Underground Research Laboratory (Japan Atomic Energy Agency, 2020). Available at: https://www.jaea.go.jp/english/index.html [Accessed 16 February 2020]. Used with permission of the Japan Atomic Energy Agency.

process will be completed by 2025, and that a geological repository will be available by 2035 (WNA, 2019).

10.10 Russian Federation

10.10.1 *Status of Nuclear Energy*

Russia is a world leader in nuclear energy. It has 35 power nuclear reactors (20 water–water energetic reactors, 10 high-power channel reactors, three light-water reactors, and two sodium-cooled breeder reactors) located in 10 power plants (WNA, 2019). In 2018, 18% of Russia's electricity was derived from nuclear energy. Rosatom, the State Atomic Energy Corporation, plans that 40–50% of the country's electricity will be produced by nuclear energy by 2050 (WNA, 2019). Six new nuclear reactors are currently under construction and 25 more are planned to be in operation by 2023 to 2035. An additional 22 power reactors have been proposed. Russia has the ability to enrich uranium, fabricate reactor fuel, chemically reprocess used fuel using the Purex process, and to solidify reprocessing waste by vitrification. A major motivation for Russia to expand its use of nuclear energy is that the country has about 9% of the world's uranium resources (see Chapter 3). ARMZ (Atom Rare Metals Gold) Uranium Holding Company is responsible for uranium mining in Russia. ARMZ operates underground mines, heap leaching and *in situ* leaching (see Chapter 3) in four geographic regions. Russia is a major exporter of uranium to the international market (WNA, 2019).

10.10.2 *Waste Management*

Radioactive waste categories in Russia resemble the IAEA system (Appendix B). OECD-NEA (2018) provided the classification system that is based on the specific activity of each waste category (Table 10.2). Rosatom and the National Operator for Radioactive Waste Management (NORWM) are responsible for the management of radioactive wastes and used nuclear fuel (WNA, 2019).

Historically the management of radioactive wastes was not a major priority in the former USSR and the early Russian Federation. Bradley (1997) provided a detailed account about how radioactive wastes were managed in the former Soviet Union. It has been estimated that there are about 1,400 near-surface burial sites (US DOE, 2011). Each nuclear facility has its own burial site. The majority of the low-level and intermediate-level wastes were buried in shallow trenches and covered with soil. Detailed information about

Table 10.2. Radioactive waste categories based on specific activity in Russia (OECD-NEA, 2018).

Waste category	Alpha + transuranic sources	Beta sources
	Bq/g	
VLLW	<100	<1,000
LLW	100–1,000	1,000–10,000
ILW	1,000–10^6	10,000–10^7
HLW	>10^6	>10^7

the disposal sites is not available. NORWM plans to establish some type of regional disposal sites for low- and intermediate-level wastes (WNA, 2019).

Russia uses deep-well injection of liquid low-level and intermediate-level wastes derived from chemical reprocessing at about 20 sites (see Chapter 6 for details). This waste disposal method relies on geological barriers above the injection zone to control the migration of subsurface radionuclides. This technology is similar to the accepted practice of deep-well injection of liquid hazardous wastes in the US. Deep-well injection is the controlled placement of liquid hazardous wastes below a confining layer into a water-permeable zone that is below drinking-water sources (see Roy *et al.*, 1991 for an example in Illinois, US). Russia plans to continue deep-well disposal of liquid radioactive wastes until at least 2030 (WNA, 2019).

It is Russian policy to chemically reprocess used nuclear fuel and to fabricate mixed oxide (MOX) fuel. However, only 16% of the used fuel available in 2011 was reprocessed (WNA, 2019). Used fuel is currently placed in wet and dry storage. A dry storage facility at Zheleznogorsk is reported to be the largest dry storage facility in the world (WNA, 2019). Russia plans to establish a deep geological repository for vitrified high-level waste. The Kola Peninsula in the far northwest area of Russia has been selected for detailed geological studies. NORWM proposed to construct an underground research laboratory by 2024 in the Pre-Cambrian-age Nizhnekansky Granite Massif. The dominate composition of the massif is biotite granite and granodiorite (Lebedev *et al.*, 2001). Based on the field data and experiences gained from the underground laboratory, a geological repository for both intermediate-level and vitrified high-level wastes may be constructed and operational by about 2035 (Lebedev *et al.*, 2001; WNA, 2019).

10.11 Kingdom of Sweden

10.11.1 *Status of Nuclear Energy*

Sweden has eight nuclear power reactors (five boiling-water reactors and three pressurized-water reactors) located in three power plants.

Table 10.3. Radioactive waste classification system used in Sweden (compiled from NWP, 2016 and OECD-NEA, 2018). SFR is the Final Repository for Short-Lived Radioactive Waste.

Waste category	Characteristics	Disposal
VLLW-SL waste	Power plant operational wastes. Decommissioning wastes. "Short" defined as a half-life of <34 years. Contact dose rate <0.5 mSv/h.	Shallow land disposal
LLW-SL waste	Contact dose rate <2 mSv/h.	SFR
ILW-SL waste	Contact dose rate <500 mSv/h.	SFR
LILW-LL waste	"Long" defined as a half-life of >31 years	SFR
HLW	Spent nuclear fuel	Clab, then a deep geological repository

Nuclear power provides about 35% of the country's electricity (WNA, 2019). The long-term future of nuclear energy in Sweden has not been decided by the Swedish government. Various political parties vacillate between a complete nuclear phase-out and building new reactors at exiting sites as replacements (WNA, 2019). Sweden does not enrich or fabricate reactor fuel and depends on importing reactor fuel from France and Russia. Sweden does not have any uranium mines and does not chemically reprocess spent nuclear fuel. Spent nuclear fuel is classified at a high-level waste (Table 10.3).

10.11.2 *Waste Management*

Sweden is a world leader in managing radioactive wastes from generating electricity. The radioactive waste categories used in Sweden are similar to those outlined by the IAEA (Appendix B). The classification system used in Sweden was established by the nuclear industry (Table 10.3). There is also an Exempt or Cleared Material category that does not require a repository (NWP, 2016).

Figure 10.10. Very low-level short-lived waste burial at the Oskarshamn Nuclear Power Plant (MES, 2017). Used with permission of Oskarshamns Kraftgrupp AB.

Svensk Kärnbränslehantering AB (Swedish Nuclear Fuel and Waste Management Co, SKB) is responsible for managing and disposing of the radioactive wastes from the Swedish nuclear power plants including spent nuclear fuel. The nuclear power plants at Righals, Forsmark, and Oskarshamn and the Studsvik Nuclear Complex dispose of solid very low-level short-lived waste (VLLW-SL) on site. The three powerplants bury VLLW-SL above ground after placing the containers on a concrete pad (Fig. 10.10) or on bedrock (OECD-NEA, 2018). All other low-level and intermediate-level wastes are disposed in a geological repository called *Slutförvar för kortlivat radioaktivt avfall* (SRF) which is referred to as the Final Storage for Reactor Wastes or the Final Repository for Short-Lived Radioactive Waste. The SRF is discussed in detail in Chapter 6.

All Swedish spent nuclear fuel has been stored at an interim repository at Oskarshamn since 1985. The spent fuel is stored under 8 m of water in one of two pools in an excavation that is 30 m below the surface. The interim facility is called *Centralt mellanlayer för använt Kärbränsle* or Clab. It is referred to as the Central holding storage for spent nuclear fuel or the Central Interim Storage Facility for Spent Nuclear Fuel. Additional information about Clab is given in Chapter 6.

Sweden plans to construct a deep geological repository for spent nuclear fuel. The spent fuel at Clab will be transported to the repository when it is available. The Äspö Hard Rock Laboratory (HRL) was constructed as an underground research laboratory to conduct research and development needed for the future repository. The Äspö HRL was constructed below Äspö Island from 1990 to 1995. It consists of a main tunnel that spirals downward to a depth of 460 m (Fig. 10.11). The total length of the tunnel is 3,600 m (Stenberg and Wikberg, 2017). The geology of the site consists of Pre-Cambrian-age granite, granodiorite, diorite and gabbro. A detailed description

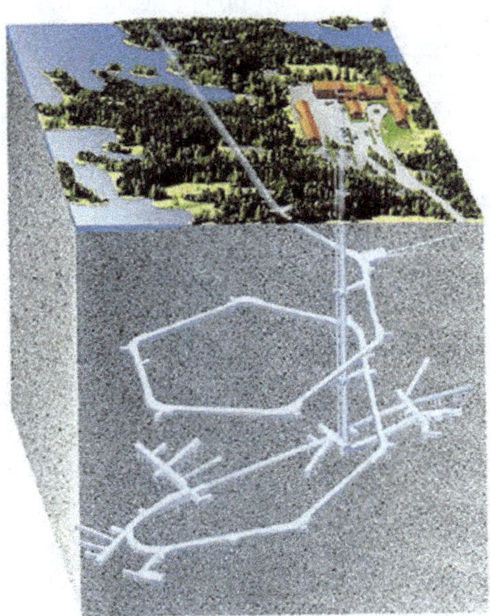

Figure 10.11. The Äspö HRL. The levels created for field-scale experiments are concentrated at the 400, 410 and 450 m depths. Used with permission of the Royal Institute of Technology.

Table 10.4. Partial summary of the investigations and activities conducted at the Äspö HRL (from Stenberg and Wikberg, 2017).

Geochemical investigations
Radionuclide transport processes Sorption and diffusion of radionuclides Impacts of oxidization–reduction reactions on radionuclides Impacts of oxygen depletion and sulfide production
Hydrogeological investigations
Flow paths of groundwater within fractures Groundwater tracer tests
Repository design, construction and performance
Behavior of the bentonite seals and backfill when in contact with groundwater, gases, elevated temperatures, and bacterial growth. Geotechnical rock mechanics Placement and retrieval of the SKB-3 copper canisters Field-scale experience, training, and presenting information to technical audiences and to the general public

of the geology and stratigraphy of Äspö Island was provided by Wahlgren (2010).

Several research projects have been conducted at the Äspö HRL. Stenberg and Wikberg (2017) provided a summary of the numerous investigations conducted to collect research data and to gain field-scale experience necessary for the construction of the repository (Table 10.4). A major component in SKB's research program was developing and testing the KBS-3 Method for the placement and retrieval of copper canisters holding the spent nuclear fuel (see Chapter 6).

After a detailed site-selection process, SKB selected the Östhammar municipality with local consent for the construction of the geological repository on the basis that the site had "the best geology." The planned depth of disposal is about 500 m, and the subsurface geology was drier and less fractured than that found at the Äspö HRL. The application to begin constructing the repository was sent to the Swedish Radiation Safety Authority and the Land and Environmental Court in Stockholm in 2011. However, the Land and Environmental Court requested additional information about the copper canisters, and research is on-going. SKB plans to begin construction of the repository in the early 2020s.

10.12 United Kingdom of Great Britain and Northern Ireland

10.12.1 *Status of Nuclear Energy*

The United Kingdom (UK) has 15 nuclear power reactors (14 advanced gas-cooled reactors and one pressurized-water reactor) located in eight power plants. Six power plants are located in England and two are in Scotland. The 15 reactors currently generate about 21% of the UK's electricity (WNA, 2019). The current UK energy policy is to reduce carbon dioxide emissions from fossil fuels. The role of nuclear energy in the UK in the future appears to be certain. One new pressurized-water reactor is under construction. Nine new reactors have either been planned or proposed although construction dates are unavailable. All of the future reactors are planned for existing or former power plant sites (WNA, 2019).

The UK has the capacity for uranium fuel conversion, enrichment, fuel fabrication, chemical reprocessing of used fuel and waste treatment of reprocessing wastes by vitrification. Most of the facilities are located at Sellafield which is the UK's major nuclear complex for legacy wastes from the Cold War. The UK has recently decided to cease reprocessing used fuel with the option of resuming the process at a later date (WNA, 2019). The UK has some minor uranium deposits, mostly notably in south-west England (BGS, 2010). There are, however, no active uranium mines in the UK. Uranium is imported mostly from Australia.

10.12.2 *Waste Management*

The radioactive waste categories used in the UK are not strictly based on the IAEA system (Appendix B). The criteria are based on

Table 10.5. Radioactive waste classification used in the UK (NWP, 2016).

	Spent fuel
	High-Level Waste. Also called Higher Activity Waste (HAW) Significant decay heat
↑	Intermediate-Level Waste (also HAW) >4 GBq/tonne alpha sources and/or >12 GBq/tonne beta and gamma sources
Activity	Low-Level Waste <4 GBq/tonne alpha sources and <12 GBq/tonne beta and gamma sources
	Very Low-Level Waste (high volume and low volume)
	Out-of-Scope (not subject to regulatory control)

radionuclide activity but not on the basis of half-life (Table 10.5). The Nuclear Decommissioning Authority (NDA) is a government agency within the Department of Business, Energy, and Industrial Strategy. The NDA is responsible for radioactive waste management. Radioactive Waste Management (RWM) Limited is a subsidiary of the NDA and is devoted to managing high-level wastes and spent nuclear fuel.

All solid low-level radioactive wastes are disposed at the UK's national facility called the Low Level Waste Repository at Drigg in Cumbria (Bergström *et al.*, 2011). The Drigg facility has been in operation since 1959 and was initially a landfill. Radioactive wastes in drums, bags, and loose debris were dumped into shallow trenches then covered with about 1.5 m of soil. The design of the repository has evolved since about 1988. Surface concrete vaults have been erected on a concrete base (Fig. 10.12). Low-level wastes are now compacted prior to placement in 20-m^3 steel drums which are grouted when full. The Drigg repository receives low-level wastes from national fuel cycle facilities, industrial sources, hospitals, universities, and site-remediation projects (Bergström *et al.*, 2011). The concrete vaults will be covered when the facility is closed.

Vitrified high-level wastes from reprocessing are stored in stainless steel containers at the Waste Vitrification Plant at Sellafield

Figure 10.12. Vault number 9 at the Low Level Waste Repository at Drigg in Cumbria (LLW Repository Ltd, 2020]. Available at: https://www.gov.uk/government/organisations/low-level-waste-repository-ltd [Accessed 17 February 2020]. Used with permission of the Nuclear Decommissioning Authority.

(WNA, 2019). Spent nuclear fuel is managed by both wet storage at the source power plants, and in dry storage casks. For example, a HI-STORM cask was recently placed inside a new building at the Sizewell B Power Station (see Hambley *et al.*, 2016). The RWM plans to construct a deep geological repository for spent fuel, high-level wastes, and some types of intermediate-level wastes. The site-selection process, however, has not been completed. Consequently, the geological conditions of potential sites are not known. Unlike the other countries discussed in this chapter, there appears to be no plans to construct an underground research laboratory prior to the repository (Tweed, 2016). The geological repository may be available for wastes by about 2040 (WNA, 2019). The Scottish government, however, does not support the development of a geological repository (Tweed, 2016). The government's policy is to manage higher-activity wastes in "near-surface facilities" but provided no details or timelines (Scottish Government, 2011).

10.13 Review Questions

1. In the radioactive waste classification system used by some of the countries, how is the term "short half-life" defined?
2. What is the purpose of an underground research laboratory?
3. Of the 12 countries discussed in this chapter, what type of geological materials are going to be used to construct a waste repository?
4. Which of the 12 countries have used former mines as radioactive waste repositories?
5. Which of the 12 countries are likely to have an operational geological repository before 2035?
6. Which of the 12 countries must import uranium or uranium fuel to sustain their nuclear energy program?
7. Which of the 12 countries currently do not chemically reprocess spent nuclear fuel?
8. Describe and contrast the various facilities used in the 12 countries to manage solid low-level wastes.
9. Why is bentonite clay a common material used in the design of most geological repositories?
10. Define the following acronyms and abbreviations used in this chapter:

    NDA SKB ONKALO MACSTOR HADES NUMO

Bibliography

Aertsens, M., Maes, N., Van Ravestyn, L., and Brassinnes, S. (2013). Overview of Radionuclide Migration Experiments in the HADES Underground Research Facility at Mol (Belgium). *Clay Minerals*, 48, pp. 153–166.

Bergström, U., Pers, K., and Almén, Y. (2011). *International Perspective on Repositories For Low Level Waste*. Swedish Nuclear Fuel and Waste Management Company (Report number SKB R-11-16).

BGS. (2010). *Uranium*. British Geological Survey. Mineral Profile.

Bradley, D. J. (1997). *Behind the Nuclear Curtain: Radioactive Waste Management in the Former Soviet Union*. Battelle Press, Columbus, Ohio.

Bräuer, V. (2016). Current Status of Nuclear Waste Disposal in Germany., In Faybishenko, B., Birkholzer, J., Sassani, D. and Swift, P. (eds.), *International Approaches for Deep Geological Disposal of Nuclear Waste: Geological Challenges in Radioactive Waste Isolation. Fifth Worldwide Review*, Chapter 9,

Lawrence Berkeley National Laboratory (Report number LBNL-1006984), pp. 9-1–9-16.

CRNR. (2005). *Joint Convention on the Safety of Spent Fuel Management and on the Safety of Radioactive Waste Management*, Czech Republic National Report.

Deguchi, A., Umeki, H., Ueda, H., Miyamoto, Y., Shibata, M., Naito, M., and Tanaka, T. (2016). Progress in the Geological Disposal Program in Japan., In Faybishenko, B., Birkholzer, J., Sassani, D. and Swift, P. (eds.), *International Approaches for Deep Geological Disposal of Nuclear Waste: Geological Challenges in Radioactive Waste Isolation. Fifth Worldwide Review*, Chapter 12, Lawrence Berkeley National Laboratory (Report number LBNL-1006984), pp. 12-1–12-22.

Dutzer, M., Vervialle, J. P., and Charton, P. (2006). Present Issues for Centre de la Manche Disposal Facility. TOPSEAL. In *Transactions International Topical Meeting, Olkiluoto Information Center*, Finland, September 17–20, 2006, pp. 73–77.

Farin, S., Labalette, T., Ouzounian, G., and Plas, F. (2016). Progress Towards Geological Disposal of High-Level and Intermediate-Level Long-Lived Radioactive Waste at an Industrial Scale: Th.), *International Approaches for Deep Geological Disposal of Nuclear Waste: Geological Challenges in Radioactive Waste Isolation. Fifth Worldwide Review.* Chapter 8, Lawrence Berkeley National Laboratory (Report number LBNL-1006984), pp. 8-1–8-25.

Fennovoima. (2016). *Environmental Impact Assessment Program for Spent Nuclear Fuel Encapsulation Plant and Final Disposal Facility.* Poyry Finland Oy (Project number 101001087).

Filho, P. F. L. H., Guerrero, J. P., Heilbron, M. C. P. L., Valeriano, C. M., and Silva, C. (2016). Radioactive Waste Management in Brazil Including Spent Fuel. In Faybishenko, B., Birkholzer, J., Sassani, D. and Swift, P. (eds.), *International Approaches for Deep Geological Disposal of Nuclear Waste: Geological Challenges in Radioactive Waste Isolation. Fifth Worldwide Review,* Chapter 2, Lawrence Berkeley National Laboratory (Report number LBNL-1006984), pp. 2-1–2-25.

FNCA. (2007). *Updated Consolidated Report on Radioactive Waste Management in FNCA Countries.* Forum for Nuclear Cooperation in Asia (Report number FNCA RWM-R004).

Hambley, D., Laferrere, A., Walters, W. S., Hodgson, Z., Wickham, S., and Richardson, P. (2016). Lessons Learned from a Review of International Approaches to Spent Fuel Management. *EPJ* (*European Physical Journal*) *Nuclear Sciences and Technologies*, 2, pp. 1–7.

Haverkamp, B., Biurrun, E., and Kucerka, M. (2005). Update of the Safety Assessment of the Underground Richard Repository, Litomerice. In *Waste Management Conference*, Tucson, Arizona, February 27 to March 3, 2005.

Huys, N., Braeckeveldt, M., and Ghys, B. (2016). 40 Years of Experience of NIRAS/Belgoprocess on The Interim Storage of Low-Intermediate and High-Level Waste. In *International Conference on the Safety of Radioactive Waste*

Management. International Atomic Energy Agency, November 21–25, 2016, Vienna, Austria (Report number IAEA-CN-242).

Jensen, M., Facella, J. A., Gierszewski, P., Hatton, C., and Russell, S. (2016). Progress Towards Long-Term Management of Used Nuclear Fuel in Canada. In Faybishenko, B., Birkholzer, J., Sassani, D. and Swift, P. (eds.), *International Approaches for Deep Geological Disposal of Nuclear Waste: Geological Challenges in Radioactive Waste Isolation. Fifth Worldwide Review*, Chapter 4, Lawrence Berkeley National Laboratory (Report number LBNL-1006984), pp. 4-1–4-14.

Lebedev, V. A., Akhunov, V. D., Lopatin, V. V., Kamnev, E. N., and Rybalchenko, A. I. (2001). Disposal of Radioactive Waste in Deep Geological Formations in Russia: Results and Prospects. In Withspoon, P. A. and Bodvarsson, G. S. (eds.), *Geological Challenges in Radioactive Waste Isolation*, Chapter 23. Lawrence Berkeley National Laboratory (Report number LBNL-49767), pp. 219–224.

MES. (2017). *Sweden's Sixth National Report Under the Joint Convention on the Safety Of Spent Fuel Management and on the Safety of Radioactive Waste Management.* Ministry of the Environment Sweden (Report number Ds 2017:51).

MESDE. (2015). *Energy Transition for Green Growth Act: User Guide for the Act and Its Attendant Actions.* Ministry of Ecology, Sustainable Development and Energy (France).

Neerdael, B. and Vokaert, G. (2001). The Belgium RD & D Program on Long-Lived and High-Level Waste Disposal: Status and Trends. In Withspoon, P. A. and Bodvarsson, G. S. (eds.), *Geological Challenges in Radioactive Waste Isolation,* Chapter 5. Lawrence Berkeley National Laboratory (Report number LBNL-49767), pp. 47–54.

NWP. (2016). *International Approaches to Radioactive Waste Classification.* LLW Repository Limited National Programme Office (United Kingdom) (Report number NWP-REP-134).

OECD-NEA. (2018). Radioactive Waste Management Programmes in OECD/NEA Member Countries. Organisation for Economic Co-operation and Development-Nuclear Energy Agency. Available at: https://www.oecd-nea.org/rwm/profiles/ [Accessed 17 February 2020].

Osawa, H., Koide, K., Sasao, E., Iwatsuki, T., Saegusa, H., Hama, K., and Sato, T. (2015). Current Status of R & R Activities and Future Plan of Mizunami Underground Research Laboratory. In *15th International High-Level Radioactive Waste Management Conference, Charleston, SC*, April 12–16, 2015, pp. 371–378.

Roy, W. R., Seyler, B., Steele, J. D., Mravik, S. C., Moore, D. M., Krapac, I. G., Peden, J. M., and Griffin, R. A. (1991). Geochemical Transformations and Modeling of Two Deep-Well Injected Hazardous Wastes. *Ground Water*, 29, pp. 671–677.

SNR. (2017). *Sixth National Report on Compliance with the Joint Convention Obligations.* Joint Convention on the Safety of the Management of Spent Fuel and on the Safety of the Management of Radioactive waste (France).

Stenberg, L. and Wikberg, P. (2017). The Äspö Hard Rock Laboratory. In Claesson, T., Gudowski, W., Maskenskaya, O., Roy, W., Stenberg, L., Wikberg, P. and Evins, L. Z., *Geological Storage of Spent Nuclear Fuel.* Royal Institute of Technology, Stockholm, Sweden, Chapter 12, pp. 109–149.

Schneider, M. and Marignac, Y. (2008). *Spent Nuclear Fuel Reprocessing in France*. International Panel on Fissile Materials, Princeton, New Jersey (Research Report No. 4).

Scottish Government. (2011). *Scotland's Higher Activity Radioactive Waste Policy 2011* (Report number DPPAS11098 (01/11)).

Shu, Y., Liu, Z., Lin, X., and Wang, R. (2016). A review of the development of nuclear waste treatment for China's nuclear power industry. Advances in Engineering Research, 94. Available at: https://www.atlantis-press.com [Accessed 17 February 2020].

Slovak, J. and Woller, F. (2016). Progress of the Czech Deep Geological Repository Program in Faybishenko, B., Birkholzer, J., Sassani, D. and Swift, P. (eds). *International Approaches for Deep Geological Disposal of Nuclear Waste: Geological Challenges in Radioactive Waste Isolation. Fifth Worldwide Review.* Chapter 6, Lawrence Berkeley National Laboratory (Report number LBNL-1006984), pp. 6-1 to 6-16.

Sombret, C. (1993). The Vitrification of high-level wastes in France: from the Lab to Industrial plants. *Proceedings of the CSNI Symposium on the Safety of the Nuclear Fuel Cycle. Belgium Nuclear Society, Brussels, June 3–4, 2006*, pp. 225–234.

Tweed, C. (2016). The Status of Geological Disposal in the United Kingdom Program in Faybishenko, B., Birkholzer, J., Sassani, D. and Swift, P. (eds). *International Approaches for Deep Geological Disposal of Nuclear Waste: Geological Challenges in Radioactive Waste Isolation. Fifth Worldwide Review.* Chapter 22, Lawrence Berkeley National Laboratory (Report number LBNL-1006984), pp. 22-1 to 22-18.

US DOE. (2011). *International Low Level Waste Disposal Practices and Facilities.* U.S. Department of Energy (Report number ANL-FCT-324).

Valery, J. F., Domingo, X., Landau, P. Launey, M., Deschamps, P., Pechard, C., Laloy, V., and Kalifa, M. (2015). Overview of RRSF reprocessing, from spent fuel transportation to vitrified residues storage. *RERTR 2015 36th International Meeting on Reduced Enrichment for Research and Test Reactors.* Seoul, South Korea, Oct. 11–14, 2015, pp. 1–12.

Vandenberghe, N., De Craen, M., and Wouters, L. (2014). *The Boom Clay Geology from Sedimentation to Present-Day Occurrence: A Review.* Memoirs of the Geological Survey of Belgium (Number 60-2014).

Wahlgren, C.-H. (2010). *Oskarshamn site investigation. Bedrock geology — overview and guide.* Geological Survey of Sweden (SKB R-10-05).

Wang, J., Su, R., Chen, L., Zhao, X., Liu, Y., and Zong, Z. (2016). Geological Disposal Program for High Level Radioactive Waste and the Plan for the Underground Research Laboratory in China, in Faybishenko, B., Birkholzer, J., Sassani, D. and Swift, P. (eds). *International Approaches for Deep*

Geological Disposal of Nuclear Waste: Geological Challenges in Radioactive Waste Isolation. Fifth Worldwide Review. Chapter 5, Lawrence Berkeley National Laboratory (Report number LBNL-1006984), pp. 5-1 to 5-26.

WNR. (2019). *Country Profiles.* World Nuclear Association. Available at: http://www.world-nuclear.org/ [Accessed 17 February 2020].

Appendix A

Glossary of Technical Terms

Alluvium/alluvial: A deposit of clay, silt, sand, and gravel left by flowing streams in a river valley or delta.

Aquifer: An underground layer of water-permeable rock or unconsolidated materials that yield enough groundwater to support a water well. Basically, a drinking-water source.

Bentonite: An aluminosilicate clay that often forms by the weathering of volcanic ash. Sodium bentonite swells when it comes in contact with water. Because of its swelling properties, sodium bentonite is used as a sealant or barrier between radioactive waste packages and groundwater.

Bitumen: A viscous mixture of hydrocarbons. It occurs naturally and is also made from by-products of petroleum. It is also called asphalt. It is used as a sealant in treating low-level wastes packages to make them resistant to groundwater leaching.

Cambrian: A geological time period that began about 541 million years ago and lasted for about 56 million years. The Cambrian period is characterized by the first wide-spread proliferation of marine life that yielded fossil remains.

Conversion of uranium: A step in the production of uranium fuel for reactors in which uranium oxide ore is converted to uranium fluoride for the next step of uranium-235 enrichment.

Diorite: An igneous rock type that is often referred to as "salt and pepper" because of the presence of white-colored and dark-colored minerals (hornblende and mica).

Enrichment of uranium: A step in the production of uranium fuel for reactors in which the amount of uranium-235 in the feed uranium is increased or enriched to an amount greater than that which occurs naturally (0.7%).

Erg: A unit of energy or work equal to 10^{-7} joules.

Fabrication: The last step in the production of uranium fuel for reactors in which the feed uranium is converted into ceramic uranium oxide pellets, then loaded into zirconium alloy tubes that are bundled into a fuel assembly.

Fissile: In nuclear energy, a fissile material is one that can sustain a nuclear fission-chain reaction such as uranium-235.

Gabbro: An igneous rock that forms during the slow cooling of magma and is typically dark-colored and coarse-grained.

Glacial till: A heterogenous mixture of sediments, rocks, and sometimes boulders that was derived by the erosion and subsequent deposition of glacial ice. Numerous continental and mountain glaciations have occurred during geological time spreading glacial till as the ice advances.

Granite: An igneous rock derived from the cooling of magma. It is typically coarse-grained and composed of pink-, white-, and black-colored minerals. Granite is a common building material.

Granodiorite: An igneous rock derived from the cooling of magma. It is typically medium- to coarse-grained and composed of white- and black-colored minerals. It is compositionally between granite and diorite.

Gross alpha and gross beta: A screening procedure that measures radioactivity in drinking water from alpha- or beta-radiation source. The test does not identify individual radionuclides.

Illite: A common, non-expanding clay mineral that often forms from the weathering of other minerals. It was named after the State of Illinois in the US.

Joule: A unit of energy or work. One Joule is equal to $1\,\text{kg-m}^2/\text{s}^2$ when measuring kinetic energy. The unit was named after the English physicist James Joule.

Metamorphic rocks: A broad term for rock types that have been subjected to significant temperatures and pressures to transform their physical, chemical, and mineralogical composition into a new type of rock. The precursor material may be sedimentary or igneous rocks. Gneiss is a metamorphic rock mentioned in this textbook and it forms when shale is subjected to heat and pressure and is metamorphosed during geological time.

Neogene: A geological time period that began 23 million years ago and lasted 20.5 million years. Early ancestors of humans appeared in Africa near the end of this period.

Oligocene: A geological time period that started about 33.9 and ended about 23 million years before present. The Oligocene is characterized by the dawn of modern ocean circulation, and the expansion of grasslands in Europe.

Operational wastes: A term used in some countries for low-level radioactive wastes created from the routine operation of nuclear power plants.

Overburden: In this textbook, the term refers to any soils, sediments, and rock layers that are between the surface and the subsurface area of interest such as a potential geological repository.

Overpack: A secondary or additional outer container for one or more waste packages that is used for handling, transport, storage or disposal of radioactive wastes.

Pitchblend: An older name for uraninite which is dominantly composed of uranium oxide.

Pre-Cambrian: The earliest geological time period that ended about 540 million years ago with the beginning of the Cambrian Period and began with the formation of the Earth. Characterized by sparse fossil remains and a scarcity of surface exposures, this period covers about 88% of the Earth's geological history.

Raffinate wastes: A liquid by-product that results after a useful product has been chemically extracted. In this textbook, a raffinate is a complex mixtures of acids, organic solvents, fission products and other many other dissolved metals. This raffinate results from reprocessing used nuclear fuel.

Riprap layer: A riprap layer is composed of large rocks or demolition debris such as broken concrete blocks and bricks. The layer is applied to surfaces to reduce erosion of lakeshores, levees — and in this textbook — disposal cells constructed to isolate legacy wastes.

Specific activity: The radioactivity per unit mass of a radionuclide. Also, the activity per unit mass or volume of the material in which the radionuclides are essentially uniformly distributed.

Superfund site: A Superfund site is any land area in the US that has been contaminated by hazardous materials or wastes and classified by the US EPA as a candidate for cleanup because it poses a *significant* risk to human health and the environment. The US Congress established the Comprehensive Environmental Response, Compensation and Liability Act in 1980 which is now informally called Superfund.

Vitrification: A chemical process for converting a substance into a glass-like, non-crystalline amorphous solid. In this textbook, vitrification is final step in converting a liquid waste raffinate into a solidified waste form that is relatively resistant to leaching if it comes in contact with groundwater.

References Consulted to Compile this Glossary

Allaby, M. (2013). *A Dictionary of Geology and Earth Sciences*, 4th Ed., Oxford University Press, DOI: 10.1093/acref/9780199653065.001.0001

Geology Dictionary. (2019). Available at: https://geology.com/geology-dictionary.shtml [Accessed 17 February 2020].

IAEA. (2000). *Terminology Used in Nuclear, Radiation, Radioactive Waste and Transport Safety*. International Atomic Energy Agency, version 1, 153 p.

Appendix B

The International Atomic Energy Agency Classification of Radioactive Wastes

"One fact remained inexplicable — that of the compass"

— Henry Lawson in *Journey to the Center of the Earth* by Jules Verne

The International Atomic Energy Agency (IAEA, 2009) provided a classification system for radioactive wastes. The key to understanding this classification system is for the reader to understand that the IAEA provided only a guide for individual countries to adapt, and then derive quantitative criteria for waste classification. It was the goal of IAEA (2009) to "provide a framework for defining waste classes . . . and to serve as a tool for facilitating communication on radioactive waste safety." IAEA concluded that their generic system would need to be adapted in individual countries by considering acceptance criteria, long-term disposal options and policies, and the characteristics and availability of the disposal facilities in each country. With that said, six waste categories were provided for solid radioactive wastes (Fig. B.1). The text below for each category is paraphrased or directly quoted from IAEA (2009). The author also provided some interpretations and editing to make the text clearer.

1. **Exempt Waste (EW):** This waste type meets the criteria for clearance or release. The material is not considered to be a significant radioactive waste. The release criteria may vary depending on the country. IAEA (2009) suggested that an effective dose of $\leq 10\,\mu$S/year (≤ 1 mrem/year) could be a useful limit.

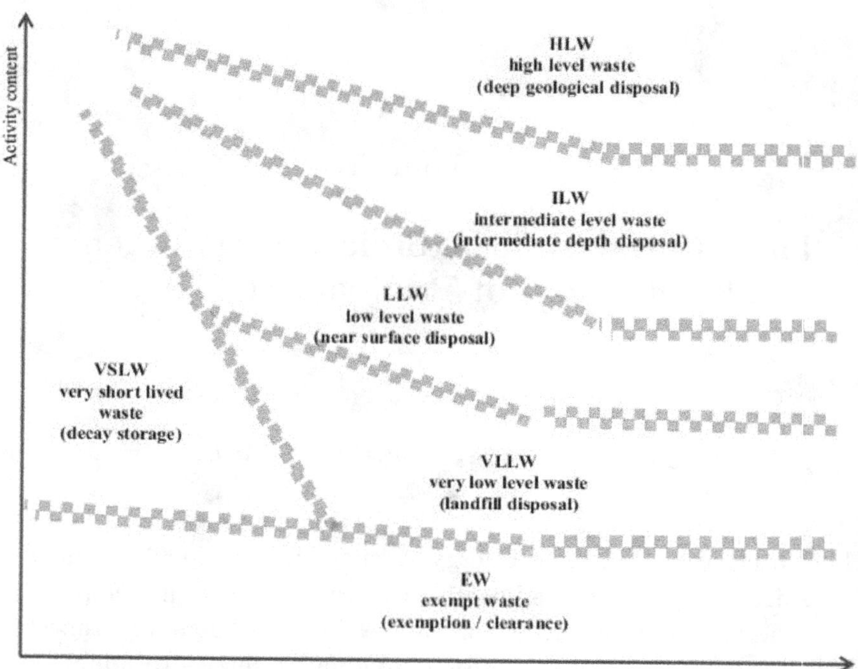

Figure B.1. IAEA radioactive waste classification categories. Note that there are no numerical criteria for either half-life or activity (IAEA, 2009). Used with permission of the International Atomic Energy Agency.

2. **Very Short-Lived Waste (VSLW):** The major criterion for this waste type is that the predominant radionuclides have a relatively short (≤100 days) half-life. This type of waste can be stored for radioactive decay for "as long as a few years." Examples include wastes from laboratory experiments and medicine. Waiting for decay prior to disposal is called Decay-in-Storage in the US (see Chapter 4).
3. **Very Low-Level Waste (VLLR):** This type of waste is not an Exempt Waste and is "suitable for disposal in near-surface landfill ... facilities." Typical sources of this type of waste include those created by powerplant operations and from the

decommissioning of nuclear facilities. Other sources may include soil and debris with "low levels of activity concentrations."

4. **Low-Level Waste (LLW):** This type of waste in not an Exempt Waste and contains "limited amounts of long-lived radionuclides." It may contain relatively large amounts of short-lived radionuclides. Short-lived is often regarded as having a half-life that is less than 30 years. This waste does not generally require shielding during "normal handling and transport." This waste is suitable for near-surface disposal for "isolation for . . . up to a few hundred years." A disposal depth of about 30 m may be required. This waste category is broadly equivalent to a Class A Waste in the US (see Chapter 4).
5. **Intermediate-Level Waste (ILW):** This type of waste "contains long-lived radionuclides in quantities that . . ." requires disposal depth between "a few tens and a few hundreds of meters." IAEA refers to this disposal scenario as "intermediate-depth disposal." IAEA argued that the boundary between LLW and ILW "cannot be specified in a general manner" because the distinction will depend on the actual disposal facility and its "supporting safety assessment." This waste category is broadly equivalent to a Class B and C Waste in the US (see Chapter 4).
6. **High-Level Waste (HLW):** This type of waste contains "such large concentrations of both short- and long-lived radionuclides [that they require] deep geological disposal with engineered barriers." HLW generate significant amounts of decay heat for centuries. This type of wastes includes spent nuclear fuel and the wastes created by reprocessing spent fuel. The subject of geological repositories for high-level wastes is discussed in Chapter 6.

As shown in Fig. B.1, a given half-life, the concentration of the radionuclide will have a greater impact on waste classification. For example, the half-life of cesium-137 is 30.2 years. Depending on its concentration in the waste, the material could be classified as EW, LLW, ILW, or HLW which all require different management techniques. IAEA argued that their classification system was designed with the goal of *long-term disposal*. For example, a waste classified initially as VSLW could become an EW waste and released in the future. It is beyond the scope of this textbook to provide detailed waste classification criteria for every country listed in Chapter 10. If required, the reader will need to conduct country-specific research for additional information.

Bibliography

IAEA. (2009). *Classification of Radioactive Waste.* International Atomic Energy Agency. General Safety Guide Number GSG-1.

Index

www.ingramcontent.com/pod-product-compliance
Lightning Source LLC
Chambersburg PA
CBHW061625220326
41598CB00026BA/3878

9789811230080